IRREVERSIBLE

WHAT CAN WE DO?

CRAIG B. SMITH
WILLIAM D. FLETCHER

Publish Authority

To request permission, contact the publisher at:
permissions@publishauthority.com

Editor: Janie Mills
Cover Design: Raeghan Rebstock
Interior Design: Teresa Evans

ISBN: 978-1-967213-15-3 (paperback)
ISBN: 978-1-967213-14-6 (e-book)

Published by Publish Authority
www.publishauthority.com

Printed in the United States of America

DEDICATION

*To the future generations who will inherit this earth.
The current generation owes them a better world
without global warming.*

CONTENTS

Figures

Tables

APPENDIX FIGURES

PREFACE:

WHY READ THIS BOOK?

Along with the threat of world war and worldwide pandemics, global warming is one of the most important and complex environmental, public policy, and international relations issues in the world today. Many aspects of human society are already being impacted in some way by global warming.

There are many signs that global warming and resulting climate changes will take a long time to reverse. But continued warming is not inevitable—if we take action promptly. Among many hopeful signs are the rapid growth of solar and wind for electricity production, increasing use of electric vehicles and projections that show the use of coal, oil, and natural gas is starting to decrease. Unfortunately, global cooperation is compromised by the U.S. withdrawal from the Paris agreement.

Stopping global warming and adapting to its unavoidable consequences is not a technical problem. It primarily requires that humans eliminate the use of fossil fuels. For the most part, the science and technology available today are sufficient to accomplish this. For example, solar and wind power are already cheaper than coal and natural gas for electricity generation in most locations. The big challenge is educating the public so that they will support the required changes, getting governments to support changing our energy economy and other actions and to achieve the level of international cooperation and coordination required.

This book provides a complete understanding of global warming and climate change. For non-technical readers, we designed the book to be an understandable and comprehensive presentation of the science and technology needed to understand global warming and what can be done about it. We describe its complexities and challenges. The book will interest the layperson concerned about global warming as well as students in technical disciplines, political science, international relations, among others.

To the best of our ability, we based the book strictly on facts that we have documented and can be verified. We have tried to keep it free of personal opinions, speculation and political influences. We leave it to readers to reach their own conclusions based upon the facts presented.

Many changes have occurred since our first book ***Reaching Net Zero: What it Takes to Solve the Global Climate Crisis*** was published in 2020. The COVID-19 pandemic caused economic disruptions and initially led to reduced greenhouse gas emissions, only to have emissions surge as the pandemic waned. In February 2022, Russia invaded Ukraine, causing fuel shortages and supply disruptions that led to global economic problems. As the war dragged on, some countries slowed their efforts to expand renewable energy, instead prioritizing finding replacement sources of fossil fuels. China and India increased their use of coal, and the US expanded production of liquefied natural gas to help Europe get through the winter.

Global warming is due to increased greenhouse gas emissions caused by human activities. Climate change is a consequence of global warming. It differs from the weather. Weather is what happens regionally over short periods of time as determined by temperature, humidity, wind speed, rain or snowfall, atmospheric pressure, and other parameters. Climate is the average atmospheric condition over relatively long periods of time, usually 30 years, for a geographic region. Climate change is a broad term that encompasses the more frequent and severe weather events we are experiencing and other effects due to the earth's higher atmospheric and ocean temperatures.

Carbon dioxide (a molecule consisting of one atom of carbon and two of oxygen and abbreviated CO_2) emissions from combustion of fossil fuels is the main cause of global warming. While CO_2 is the major greenhouse gas, amounting to about three-fourths of the total, there are other gases, such as methane, that also contribute. The combined effect of all types of greenhouse gases is often referred to as "CO_{2eq}," shorthand for "carbon dioxide equivalent."

The use of fossil fuels to produce power and heat for consumers and industry accounts for most greenhouse gas emissions. In addition, about 25 percent is due to agriculture and land use changes, including the destruction of forests to produce more land for agriculture.

In Chapter 1, we cite the scientific facts that prove global warming is real as evidenced by seven major changes in the environment. Chapter 2 follows to show how, for all practical purposes, global warming is irreversible— it will take a century or more to naturally reduce greenhouse gases in the atmosphere after human-caused emissions are eliminated. Chapter 3 explains the causes of global warming. Chapter 4 explains the dangers of global warming and its negative effects on our environment and the population. Chapter 5 describes fossil fuel energy uses that cause global warming. Chapter 6 describes alternate sources of energy that do not produce greenhouse gases. Chapter 7 is an overview of why it is so difficult to stop global warming. Chapter 8 outlines international efforts to date. Chapter 9 describes what it will take to stop global warming and summarizes a practical action plan that could be implemented starting immediately. Fortunately, there are some positive developments already occurring, as we describe in Chapter 10, but the problem is the world is not moving fast enough or on a large enough scale. This leads into Chapter 11, which presents several scenarios developed by the scientific community to answer the question of what might happen if global warming exceeds various limits. Finally, Chapter 12 lists some suggestions that we can all consider as efforts to reduce greenhouse gas emissions more rapidly. Chapter 13 includes the authors' final thoughts on dealing with global warming. We have provided appendices and references for readers who want additional details.

We are all challenged by the large amount of climate disinformation produced by the fossil fuel industry and others, and by the complexity of much of the available valid science. Several organizations and individuals minimize the impact of global warming or promote the belief that nothing can or should be done about this problem. Others are promoting naïve and unrealistic solutions or timelines that cannot be met, such as eliminating all greenhouse gas emissions by 2050. This book counters misinformation and misunderstandings concerning global warming.

There are no easy or quick solutions. The task ahead is difficult, but doable. Nations may procrastinate. The degree of cooperation, foresight, and sacrifice needed may be beyond our current capabilities on a national or global basis. We may have to experience a severe crisis before effective actions are taken. We could see a substantial increase in the earth's temperature with severe effects on the earth's climate and local weather patterns before the world finally takes action.

This book proposes practical solutions that can be implemented now with today's technology to significantly reduce greenhouse gas emissions and put the world on a path toward lower emissions. There are positive trends that need to be accelerated. An optimistic future is possible with abundant energy to maintain rising living standards without most of today's air pollution. The negative consequences of not reducing emissions are potentially severe. We should start immediately with the technology we have. Attacking this problem now will stimulate new approaches and the development of new technologies needed to cut emissions more quickly.

We emphasize our belief that humanity will survive one way or another. But without action now, it may not be in a world we want for our grandchildren or for their children. A dire future is not inevitable. We know what to do to avoid it. But so far, many have chosen to look the other way, to pretend it is not a real problem, to hope that the status quo can go on forever.

If you want to know more about global warming and what to do about it, visit our website, **https://theglobalclimatecrisis.com.** It also contains useful information and is updated. If you like what you read in this book, inform your friends. An informed and concerned citizenry is necessary to solve this critical problem.

Craig B. Smith
Santa Barbara, CA

William D. Fletcher
Newport Beach, CA

MEET CHARLIE: ©

There has been so much conversation about carbon dioxide in the last 50 years—an atom of carbon with two atoms of oxygen—that we felt the need for introductions. It's not as though readers don't know "Charlie," as we like to call this molecule, because he's an integral part of all of us. We see each other every day and every night and usually in harmony; after all, he's very essential. If it weren't for him, the earth would be a ball of ice, and we would not be writing this.

Let's start with names, to avoid any misunderstanding. We like using the International Phonetic Alphabet; so, we named him Charlie Oscar the 2ND. If you're a mathematical type and like the shorthand better, call him CO_2, or just plain Charlie.

First, let's dispense with all bad things you hear about Charlie. It's easy to be misled by misinformation and false statements, giving you the impression that Charlie's a bad guy. That's not true. It's fair to say that pretty much every living thing depends on Charlie for their existence. So, take that for granted, and we'll let Charlie get on with his story in his own words.

HOW I WAS BORN

As it was, my birth was pretty spectacular. It occurred after that intergalactic nightmare of the Big Bang. Carbon and oxygen—essential elements for life—were not formed during the Big Bang, but later within stars through nuclear fusion. Stars, especially massive and short-lived ones, fused hydrogen, helium, and other elements, eventually creating me. When stars died in supernovas, they released these elements into space where they became building blocks for new stars and planets and new elements, including me. The rest is history.

WHY I AM SPEAKING OUT?

There is so much confusion that a clear explanation of global warming and climate change is needed. The book contains the facts readers need in order to understand global warming and what can and should be done about it. I've been added to the story to help explain some of the more complicated matters in simple terms. Throughout the book, I will add some explanations of interest to the reader and also summarize the key points of some chapters.

Let me start by emphasizing that global warming is due to increased greenhouse gas emissions caused by human use of fossil fuels. Climate change is a consequence of global warming—it's not just bad weather.

ACTION NEEDS TO BE TAKEN

So far, the cooperation, foresight, and actions needed to stop global warming seem to be beyond human capabilities on a national or global basis. Humanity may have to experience a severe crisis before it wakes up and realizes the world is in trouble.

I am here to remind you that there are practical solutions that can be implemented to reduce the amount of greenhouse gases that humans are discharging to the atmosphere.

Most humans will not be severely affected by climate change in their lifetimes—at first. Many have chosen to look the other way, to pretend it is not a real problem, that the status quo can go on forever. It can't, and it won't. Trust me on this.

Sincerely,

Charlie

1

GLOBAL WARMING IS REAL

How do we know that global warming is real? There is clear evidence. This evidence is not based on computer models or forecasts, but on hard data. Here we describe the seven principal reasons that prove global warming is real.

First: The earth's temperature is rising steadily. This global warming can be ignored, perhaps for a long time. However, it cannot be avoided. Eventually, rising temperatures, sea level rise and weather changes may force the world to do something. But by that time, global temperatures will be much higher.

Figure 1.1 shows how the global average temperature has increased following WWII, commensurate with the growth in the use of coal, oil, and later, natural gas. Sharply rising global temperature is the first indicator that global warming is real. According to Berkeley Earth's global temperature report, the Earth set a new temperature record in 2024. The earth's average temperature is now about 1.5°C or 2.7°F warmer than its preindustrial baseline.

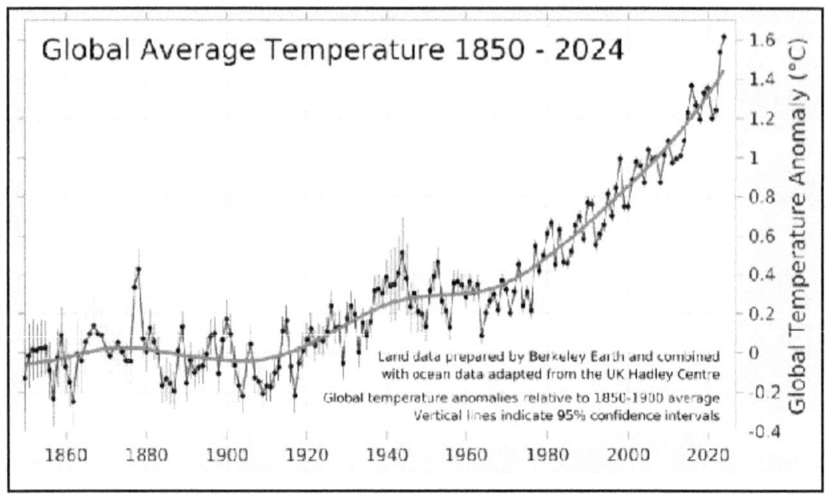

Figure 1.1 Global Average Temperature 2024 [1]

Second: Greenhouse gas concentrations in the atmosphere are easily measured, providing the second key indicator that global warming is real. In 1958, a young atmospheric scientist employed by Scripps Institution of Oceanography named Charles Keeling, began measuring carbon dioxide concentrations in the atmosphere from an observatory on the top of Mauna Loa on the Big Island of Hawaii. This site was selected because it was in the middle of the Pacific Ocean and relatively unaffected by air pollution and other effects from the continents. After several years of measurements, Keeling discovered that the concentration of carbon dioxide in the atmosphere was steadily increasing. He subsequently devoted his life to continuing these measurements and was followed later by his son and other researchers. As a result, we now know that there has been an ever-increasing concentration of carbon dioxide in the atmosphere. Other measurements indicate that the average temperature of the earth has been rising simultaneously with the increase in carbon dioxide. In effect, adding carbon dioxide to the atmosphere is equivalent to adding more layers of glass to a greenhouse.

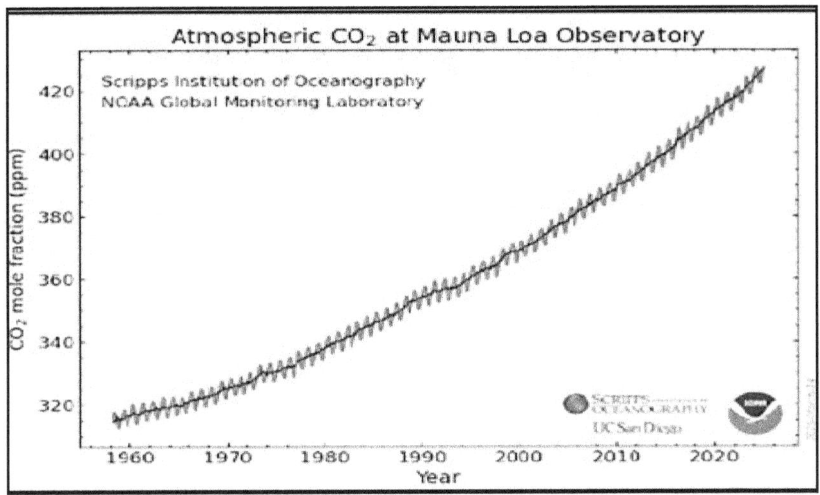

Figure 1.2 Increase of Carbon Dioxide (CO$_2$) in the Atmosphere

From preindustrial levels, the concentration of carbon dioxide in the atmosphere has risen steadily, and by May 2025 had reached 430 ppm.[2] Carbon dioxide emissions are increasing due to greater use of fossil fuels for power, heat, for manufacturing plastics and fertilizer.[3]

As greenhouse gases build up in the atmosphere, they act as an insulating layer, reducing the amount of heat leaving the earth, causing it to warm. Ocean temperatures also rise, and other physical evidence of global warming can be seen.

Third: The third indicator is that the ocean temperature is rising. Oceans absorb 90 percent of the heat created by global warming. The surface temperature of the sea is defined as the first 700 meters of depth. The temperature has increased during the 20th century and continues to rise, meaning more energy for storms, since hurricanes and typhoons absorb energy and moisture from the ocean.

Remember that one degree Celsius (⁰C) equals 1.8 degrees Fahrenheit (⁰F). Why do we have two measures for temperature?

Most of world has standardized on the metric system where temperature is measured in degrees Celsius (ºC). Reporting on global warming and climate change is written for a global audience and uses the metric system. Degrees C are hotter than ºF! However, the US and a few other small countries still use the British system measuring temperature in degrees Fahrenheit (ºF). If you are used to thinking in ºF, you might underestimate the temperature when it is in ºC. Remember that one degree ºC is equal to 1.8 ºF. Global warming today is about 1.5ºC or 2.7 ºF.

There has been a gradual increase in the average temperature of the oceans over the last 74 years. Also, as water warms, it expands, causing the sea level to rise. Sea level rise is another important indicator of global warming. (Note: additional charts and tables to this and other chapters are found in Appendix 1.)

Fourth: Rising sea levels are the fourth indicator.[4] This is due to a combination of two effects: thermal expansion of warmer ocean waters, and increased meltwater from glaciers and ice caps. To date, the melting of ice caps in Greenland and Antarctica account for about a third of total sea level rise. The average rise so far is 21 to 24 cm (8 to 9 inches) since 1880, but can be higher or lower at specific locations. Also, the rate of sea level rise is increasing.[5]

Fifth: The Arctic is heating up more than three times faster than the global average due to Arctic or polar amplification, the albedo effect, largely driven by the loss of sea ice and snow cover in the Arctic. [6,7] Snow and ice reflect most of the incoming radiation from the sun, but the exposed land and sea are much darker and absorb more of the sun's radiation. When temperatures at the poles increase faster than in the lower latitudes, the temperature differential causing ocean currents and air currents to flow toward the poles is decreased. In the northern hemisphere the jet stream becomes less stable, leading to weather changes in the northern hemisphere. [8]

Due to higher Arctic temperatures, permafrost is melting and releasing additional carbon dioxide as well as methane into the atmosphere. Arctic land areas have been a carbon sink for thousands of years, but as the area has warmed and permafrost has thawed, carbon dioxide and methane have been released into the atmosphere. [9] Also, warming trends have increased wildfires and carbon dioxide emissions north of the Arctic Circle.

Sixth: The massive ice sheets in Greenland and Antarctica are melting. In the past, these ice sheets lost ice during the summer months but replaced these losses with new snow and ice in the winter. Today, summer losses are not being fully offset. [10,11] Water from melting ice caps presently accounts for about a third of sea level rise. Greenland has been losing 267 billion metric tons per year since 2002, and Antarctica 136 billion metric tons per year.

Seventh: The acidity of the oceans has also increased over the same timeframe as the atmospheric increase of carbon dioxide, further evidence that global warming is real. About 40 percent of human-caused carbon dioxide is absorbed by the oceans. Ocean acidification, a consequence of climate change, occurs when the ocean absorbs excess carbon dioxide from the atmosphere.

There's a lot of ice out there! It's very hard to visualize how much there is. Over 99% of the world's ice is in the Antarctica and Greenland!

The Arctic is an open sea with floating ice that expands and recedes with the seasons. On average, this sea ice is about 6 to 10 feet thick. In the winter months, sea ice covers about 5.5 million square miles, about 1.7 times the area of the continental US or about the same area as all of South America. During summer months, the sea ice shrinks, covering an area only about half the size of the continental US. All this sea ice is less than one percent of the world's total ice.

Ice sheets are only found in Greenland and Antarctica. An ice sheet has two parts. An ice cap is the portion of the ice sheet sitting on land and the ice shelf is that portion of the ice sheet that is floating on seawater but that remains attached to the ice cap. Greenland does not have large ice shelves. Antarctica has five large ice shelves and several smaller ones.

Antarctica is a huge land mass, one of the seven continents, that is twice the area of the continental US. Greenland is much smaller but is still a huge island about the size of Texas or France.

The ice caps in Antarctica and Greenland are up to three miles thick. Antarctica contains about 90% of the world's ice and about 70% of the world's fresh water. Greenland holds about 8 to 9% of the world's ice and about 7% of the world's fresh water. Greenland and Antarctic ice sheets are melting slowly and contributing to sea level rise.

The Antarctic ice shelves act as dams slowing or preventing ice caps from flowing into the ocean. Warmer air and ocean temperatures slowly erode ice shelves and large parts of ice shelves have broken off in the past. If a large ice shelf collapses, it would allow a more rapid flow ice from ice caps into the ocean and increase the rate of sea level rise.

The ocean converts it into carbonic acid when it dissolves. This absorption helps regulate the planet's carbon dioxide levels but comes at a cost to marine life, particularly shellfish, corals, and other calcifying organisms.

NEED TO REACH NET ZERO

Net zero is when total greenhouse gas emissions (CO_2, CH_4 and others) are reduced such that no more emissions are added to the atmosphere. Any CO_2 emissions that can't be eliminated have to be offset by sinks such as forests that absorb excess CO_2, or by artificial means to remove carbon dioxide from the atmosphere. Achieving net zero will stabilize but not reduce the earth's temperatures until greenhouse gases already in the atmosphere gradually dissipate naturally. This could take hundreds of years.

OTHER FACTORS

There is a growing perception by the public that the weather is changing due to global warming with more severe droughts, wildfires, and storms. However, presently it is difficult to relate a specific weather event to global warming. The weather is ever changing, and it is often possible to show that a severe weather event or something similar happened in the past. In addition, there may be a human element such as building on seacoasts, or in flood plains that augment risk or building in areas where fire risks are high due to drought, presence of brush, or inappropriate building codes. Other considerations such as these examples can magnify the damage caused by more recent weather events.

It is very likely that global warming is strongly linked to heat waves and coastal flooding. There is evidence but not proof that global warming contributes to droughts, forest fires, hurricanes, and flooding. It is known that hurricanes, for example, receive their energy and their moisture from the ocean's warm surface water. Higher ocean temperatures should lead to stronger hurricanes with heavier rainfall. Deserts are expanding in some locations. Part of this is due to overgrazing.

CHAPTER SUMMARY:

Global warming is real, based on scientific measurements that reveal the following seven trends discussed above:

- Earth's temperature is rising and rapidly setting new records
- Carbon dioxide concentration is rising at an unprecedented rate
- Ocean temperatures are rising
- The sea level is rising
- The Arctic is heating up three times faster than the earth's average temperature
- The massive Antarctic and Greenland ice sheets and glaciers are melting.
- The acidity of the oceans is increasing due to dissolved CO_2

KEY POINTS OF THIS BOOK.

Table 1.1 lists key points of the book. We suggest that readers might want to make a copy of this table for use as a study guide. Also, included in the appendices is a list of recommended sources that the reader could use to access useful reports now and in the future as new information becomes available.

Table 1.1 KEY POINTS OF THIS BOOK

- Since the Industrial Revolution, humans have caused global warming, creating a threat that needs resolution without further delays.

- Global warming is irreversible. Even if human-caused greenhouse gas emissions are eliminated, the earth's temperature will remain elevated for a century or more.

- We can't get to net zero fast enough to meet the Intergovernmental Panel on Climate Change (IPCC) goal of keeping global warming under 2°C and preferably under 1.5°C. The current forecast is for warming to be about 2.5°C to 3.0°C by 2100.

- Fossil fuels are an outdated solution for human energy needs and must be replaced by renewables as quickly as possible. This will take time, and the earth's temperature will continue to rise.

- Major trends are in the right direction but are not moving fast enough, such as the growing use of renewable energy for electricity generation, electric vehicles, and electrification of industry and commerce. These efforts should be accelerated.

- There is a risk that we could reach a tipping point that will accelerate global warming in a potentially irreversible way, beyond our control. This is not a prediction but a possibility that should not be ignored.

- Due to latency, climate change will continue even when greenhouse gas emissions are reduced to net zero.

- Due to ongoing climate changes, adaptation must be a big part of any plan to deal with global warming.

- Energy austerity need not be part of the solution. A positive future is possible with abundant and affordable energy from renewable sources. This transition will take a long time.

- Fossil fuels are heavily subsidized. The biggest subsidy is the ability to discharge greenhouse gases and other pollutants into the atmosphere essentially for free. A tax on carbon emissions would speed up the transition to renewables.
- The transition to renewable energy would eliminate most air pollution in addition to greenhouse gases.
- Considering all the costs and benefits involved, funds should be available to make the transition to lower-cost renewable energy sources.
- The transition to renewable energy and changes needed to combat global warming are business opportunities for those who embrace change. China is presently way ahead of others in the production and use of solar cells, wind turbines, batteries, electric vehicles, nuclear power plants and minerals needed for renewable energy.
- If we don't do the big things, the small things may be commendable but aren't sufficient to stop global warming.
- A planned, smooth transition to renewables is needed to avoid recessions, malinvestments, stranded assets, and uninsured losses. A disruptive transition is likely to compromise public support.
- We don't have to stop global warming if we choose not to, but will have to adapt to a warmer world.

TABLE 1.1 Key Points of this Book

2

GLOBAL WARMING IS IRREVERSIBLE

As noted in Chapter 1, the earth's temperature is rising steadily due to human-caused greenhouse gas emissions. This global warming can be ignored, perhaps for a long time. But it cannot be avoided. Eventually, rising temperatures, sea level, heat, and weather changes will likely force the world to do something. By that time, global temperatures will be much higher.

We must recognize that global warming is irreversible. What does that mean? We can make the earth's temperature increase by adding greenhouse gases in the atmosphere. But we cannot remove greenhouse gases from the atmosphere in sufficient quantities to reduce the earth's temperature. Once we stop adding greenhouse gases to the atmosphere, it will take more than a century for them to dissipate naturally.

A LONG HALF-LIFE

Greenhouse gases have long half-lives and will stay in the atmosphere for a very long time even after human-caused emissions are eliminated. "Half-life" is a technical term describing the removal process. When emissions cease, the amount of each greenhouse gas will start to decline as the gas is absorbed naturally—for example, as carbon dioxide is absorbed by plants, the oceans, and soil. The half-life is how long it will take for the initial amount of gases in the atmosphere to be reduced by half.

For carbon dioxide, the most abundant greenhouse gas, it is estimated that about 60 percent will remain in the atmosphere after 50 years and about 45 percent will remain about 100 years later. It is estimated that about 20 percent of this carbon dioxide will remain in the atmosphere for much longer.

Methane, a much stronger greenhouse gas than carbon dioxide, has a shorter half-life. However, part of the methane in the atmosphere is transformed by a chemical reaction into carbon dioxide which stays in the atmosphere much longer.

HOW DID WE CREATE THIS IRREVERSIBLE SITUATION?

Mainly, it is because human use of coal, oil, and natural gas has added a huge quantity of greenhouse gases to the atmosphere. Before the rapid growth of the Industrial Revolution which began about 1850, there was an estimated 2,188 billion metric tons of carbon dioxide in the atmosphere. Carbon dioxide concentration in the atmosphere remained within the range of 280 ppm to 300 ppm over the past 850,000 years, so the amount of greenhouse gases in the atmosphere changed slowly within this limit. As mentioned earlier, following the Industrial Revolution, the amount of CO_2 and other greenhouse gases in the atmosphere began to increase.

The IPCC recommended that carbon dioxide concentration be no more than 450 ppm to limit the earth's temperature rise to 1.5°C. Charlie shows that we can only add another 157 billion metric tons of carbon dioxide to the atmosphere to stay under this limit. So, at the current rate of about 40 BMT/year, the IPCC limit will be exceeded in about four years or by 2029.

How much more carbon dioxide can we discharge into the atmosphere and stay under the IPCC recommended temperature of 1.5°C?

Charlie says the mass of carbon dioxide in the atmosphere increased and decreased over millennia. At the time of the Industrial Revolution in 1750, there was an estimated 2,188 billion metric tons (BMT) in the atmosphere. By 2025, only 275 years later, the world had added 1,175 BMT, so the total is now 3,363 BMT. How much will be there be at 450 ppm, the IPCC recommended limit? Charlie says that will be 3,520 BMT. In other words, we can only add another 157 BMT.

Reference: Fletcher, William D. and Smith, Craig B., The Global Climate Crisis: What to do about it. Elsevier, Oxford UK, 2024, pp. 305-307

IMPORTANCE OF LATENCY

In addition, we have not yet seen the full effects of greenhouse gas emissions to date due to latency, the delay between cause and effect. It takes time for a large system such as the earth to fully respond to increases in the greenhouse effect. Scientist James Hansen and others have estimated that it will take from 20 to 60 years for the full effects of emissions to date to show up.[12] If we were ever to achieve net zero—no net human-caused greenhouse gas emissions—the earth's temperature would continue to increase until the balance between the earth's heating due to the greenhouse effect is offset by its radiation of energy back into space.

ANY OTHER ALTERNATIVES?

It will never be practical, technically and economically, to remove enough carbon dioxide from the atmosphere to reduce the earth's temperature. Carbon capture is a new technology that is being developed by companies such as Carbon Engineering and Climeworks. It may have practical uses such as reducing or eliminating carbon dioxide from fossil fuel power plant exhausts and from industrial processes that cannot be converted to electricity from renewables. Carbon capture might also be used to remove carbon dioxide from the atmosphere and use it to produce synthetic fuels. However, to remove just one year's worth of human-caused emissions would cost an estimated $8.0 trillion based on Carbon Engineering's estimated cost to remove a ton of carbon dioxide from the atmosphere. A massive investment would be required to build the equipment and massive amounts of electricity would be required to power the process. See Chapter 9.

Some writers have suggested various alternatives—known generically as "geoengineering"—to offset greenhouse gas emissions. Examples included injecting sulfur dioxide crystals into the upper atmosphere to simulate a massive volcanic eruption similar to the Mount Pinatubo eruption in 1991. It produced a gas cloud that caused global temperatures to drop temporarily by about 0.5°C (0.9°F). Other proposals include scattering reflective beads into the atmosphere or covering oceans with some material to reflect heat. It is unlikely that any of these geoengineering possibilities can be implemented based upon costs and the practical difficulties of actually carrying out the proposed measure. In addition, these measures would all involve major side effects and risks.

Any substance injected into the atmosphere in sufficient quantities to affect the earth's temperature would eventually fall to earth, contaminating the land and oceans. The full effects could be unevenly distributed or have unpredictable results, such as affecting rainfall patterns.

So, again we say, global warming is irreversible.

3

WHAT CAUSES GLOBAL WARMING?

Global warming is a serious threat to life on earth as we know it and cannot be ignored. In 1824, mathematician Joseph Fourier coined the term "greenhouse effect" to describe the phenomenon thought to be responsible. Experiments led to the discovery that some gases, in particular carbon dioxide, were not transparent to infrared radiation (longer wavelengths). Sunlight is incident on the earth in the visible spectrum (light of shorter wavelength) but is reradiated back into space as infrared. This is because wavelength is a function of the temperature of the radiating body—the higher the temperature, the shorter the wavelength.

It was left to Svante Arrhenius, a Swedish Nobel Prize winner, to carry out the analysis that proved this proposition in 1896. He also calculated that doubling the amount of carbon dioxide in the atmosphere would increase the global temperature 4°C. Prophetic words indeed.

Many competent scientists—meteorologists and other scientists from all over the world—have been carefully studying this issue for more than 50 years and they come to the same conclusion.

INCIDENT SOLAR RADIATION

The source of energy for the earth is radiation from the sun. The earth is a sphere that circles the sun in an elliptical orbit once each year. This reality has important consequences for our climate. Each *day*, the earth receives an estimated 15 trillion gigajoules (GJ) of energy in the form of radiation. In 2023, the world's population was using about 640 billion GJ per *year*. So, if the incoming solar energy could be fully converted to human use, one day's worth would provide the energy needs of the earth's population for over *20 years*. It is a large amount of energy, but we receive it unevenly—more at the equator, less at the poles, and on only half of the planet at any given time.

One would think that receiving all this energy every day would cause the earth to heat up. It would, if there weren't offsetting processes at work. The incoming solar energy is short wavelength radiation that includes the visible light that we see. The visible spectrum is centered around a wavelength of 0.5 microns (19 millionths of 1 inch). The vast majority of the sun's radiant energy is not used on earth but is radiated back into space as long wavelength infrared radiation centered around a wavelength of 15 microns.

Although the earth is obviously much cooler than the sun, it is much warmer than the empty space that surrounds it. Earth's average temperature at present is estimated to be 15.2°C (59.4°F) while the temperature of space is minus 270°C or (minus 454°F), a difference of 286°C.

Over the eons, the world has been hotter and much cooler than it is today. There have been ice ages and times when there was very little ice, when the sea level rose hundreds of feet, and the land turned tropical. However, these natural changes to the earth's climate took a long time—tens of thousands of years per cycle. How do we explain these large changes in the earth's climate?

Changes in the earth's climate from warm periods to ice ages are due to changes in the amount of radiant energy the earth absorbs from the sun. There are three major perturbations to the earth's orbit around the sun. They are known as the Milankovitch cycles.[13] They have an important bearing on climate. The first is *precession*. The earth spins with a slight wobble. It completes one complete wobble roughly every 26,000 years. The earth's axis is tilted, currently about 23.5° from the vertical. This *tilt* is of great importance since it causes the seasons. The tilt fluctuates back and forth a few degrees, taking about 41,000 years to fully complete a cycle. Finally, the earth's orbit also varies *obliquely*, from elliptical to more circular, with a cycle that takes about 100,000 years. The combination of these three effects causes the earth to heat up or cool down. This is a very slow process that takes about 100,000 years for a full cycle as shown in Figure 3.1. This is very different from the much more rapid changes in the earth's temperature that have occurred in the last 150 years or so.

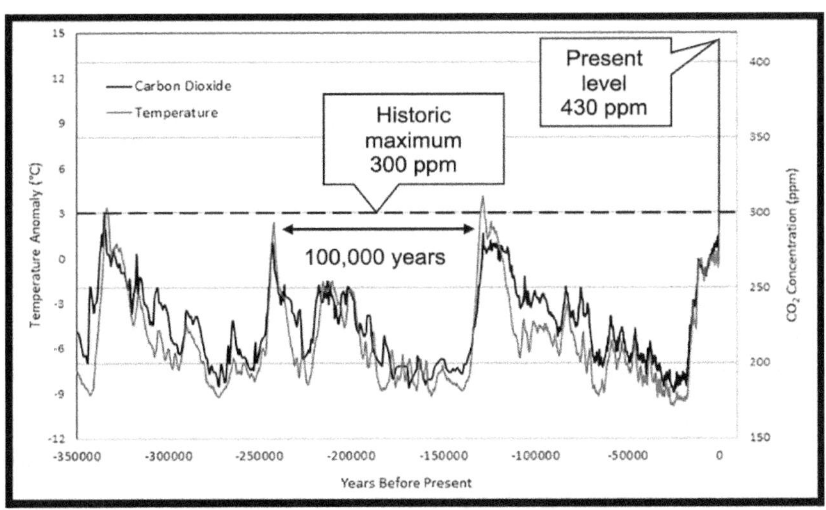

Figure 3.1: CO2 and the Earth's Temperature Gradually Change in Tandem [14]

We can also see from Figure 3.1 that the concentration of carbon dioxide in the atmosphere has been closely followed by changes in the earth's temperature. A decrease in CO_2 levels made once tropical zones cold. This close relationship has held for at least the last 850,000 years, which is as far back as sophisticated ice core measuring techniques have enabled us to look. This is an example of the earth being a dynamic (changing) system, but stable, cycling within long-established boundaries and maintaining relationships between variables, in this case carbon dioxide concentration and temperature. The earth has had many cycles of climate change in its history—usually over periods of hundreds of thousands of years. What makes these recent changes noteworthy (and ominous) is that they are occurring at an unprecedented rate—in decades, not in millennia, like in prehistoric times.

THE GREENHOUSE EFFECT

Most readers are familiar with greenhouses. The phenomenon is what we experience when an automobile is parked outside on a sunny day. Radiant energy from the sun penetrates the windows and warms the interior of the vehicle. Some of this radiant energy is absorbed by the automobile's interior. The rest attempts to leave the vehicle as radiant energy but is reflected back by the window glass, heating the interior.

Today, the earth's temperature is no longer primarily driven by the Milankovitch cycles, but by the greenhouse effect. What we are seeing is that human-caused (anthropogenic) changes are increasing the greenhouse effect and causing near-term increases in the earth's temperature. Human activity, mainly the use of fossil fuels, is overriding natural forces, leading to rapid increases in the earth's temperature. Today, human activity has entirely reversed that trend, now pushing us into dangerous warming. Temperature increases that usually take thousands of years are now happening in decades and are moving in the opposite direction from what natural cycles would dictate.

The concentration of carbon dioxide in the atmosphere did not exceed about 300 ppm until the start of the Industrial Revolution. In May 2025, the peak carbon dioxide concentration in the atmosphere reached 430 ppm, about 40 percent higher than preindustrial revolution times and rising. In addition, the earth's average temperature has risen in proportion to the increasing carbon dioxide concentration. This increase in carbon dioxide in the atmosphere and the rapid rise in the earth's average temperature are occurring too rapidly to be caused by the Milankovitch cycles. There are no other natural forces that can explain these increases.

But the greenhouse effect is essential for life on earth. The carbon dioxide in the atmosphere traps enough of the sun's incoming energy to keep the planet sufficiently warm to support life, with an average earth temperature of 13.7°C (56.7°F) during preindustrial revolution times. Without some naturally occurring greenhouse gases, the earth's average temperature would be a chilly (-)18°C (0.4°F) below freezing.[15] But this is a delicate balance, a balance that has been upset to the point that the earth is now warming. The Earth's average temperature is now about 15°C (59°F).

WHAT ARE THE GREENHOUSE GASES?

Carbon dioxide is the most common greenhouse gas and accounts for about three-fourths of greenhouse gases. The next largest greenhouse gas is methane (18 percent), followed by nitrous oxides and fluorinated gases released by industrial processes. Most of the carbon dioxide, 90 percent, comes from the burning of fossil fuels, namely coal, oil, and natural gas. Table 3.1 lists greenhouse gas types and percentage of composition.

Greenhouse Gas	Major Sources	Approximate percent
Carbon dioxide (CO_2)	Fossil fuel combustion (coal, oil, natural gas), deforestation, land use changes, cement and steel production.	75
Methane (CH_4)	Livestock (enteric fermentation), rice paddies, landfills, coal mining, oil & gas operations, wetlands.	17
Nitrous oxide (N_2O)	Agricultural soils (fertilizer use), manure management, fossil fuel combustion, industrial processes.	6
Fluorinated gases (HFCs, PFCs, SF_6, NF_3)	Refrigeration, air-conditioning, industrial processes, electronics manufacturing.	2
Water vapor (H_2O)	Naturally occurring (evaporation from oceans, lakes), not directly emitted by human activities, but influenced indirectly via warming.	variable

Table 3.1: Types of Greenhouse Gases

WHERE DO GREENHOUSE GASES COME FROM?

In 2023, the estimated total global emission of greenhouse gases was 57 billion metric tons (GMT) of CO_2eq. This was the largest annual level ever recorded. Table 3.2 lists the primary sources responsible for these emissions. [16]

Source	Percent
Production of power and heat	68
Various manufacturing processes	9
Agriculture and land use changes	18
Waste and miscellaneous	5
TOTAL	100

Table 3.2: Sources of Greenhouse Gases

In the previous section, we summarized emissions by greenhouse gas type. Table 3.2 is a summary of total global greenhouse gas emissions by source where they came from. Energy use is the largest single source of greenhouse gases—about 68 percent. The biggest use of energy is for power including the production of electricity. Transportation is the second largest energy use. As stated earlier, the efficiency of power plants and internal combustion engines is about 30 to 40 percent so that 60 to 70 percent of the energy used in these sources is waste heat discharged into the atmosphere through cooling water, smokestacks, radiators or engine exhausts. Various industrial processes are responsible for about 9 percent of emissions. Cement, the world's most widely used building material, accounts for about a third of these emissions. Agriculture, forestry, and other land use changes (AFOUL) account for about 18 percent of emissions. This includes a broad category of land use, land use change, and forestry (LULUCF) which is largely the destruction of forests that reduce nature's ability to absorb carbon dioxide from the atmosphere. The final category is various sources of waste disposal practices that release greenhouse gases. Refer to Figure 6 in Appendix 1 for a detailed breakdown of about 20 different sources.

Today, fossil fuels produce 81 percent of the world's energy, nuclear energy about 5 percent, while the remaining 14 percent is produced by renewable energy forms, namely wind, solar, and hydropower. Using renewables to produce electricity is the fastest growing source of energy in the world, accounting for over 90 percent of new power projects in 2024. The cost of electricity from solar panels and wind turbines has decreased significantly in the last 20 years as cost and efficiency continue to improve. This makes electricity from renewables the cheapest source of energy, compared to fossil fuels and nuclear power in an increasing number of locations. About 30 percent of electricity now comes from renewable sources—more than is generated by coal.[17]

For energy production, the use of coal and natural gas to produce electricity is the single largest use of fossil fuels but is being replaced by renewables as their cost continues to decline. China, India and others are building new coal-fired generating stations. The next largest use of fossil fuels is for transportation, followed by industry, residential and commercial use, and agriculture. Other land use changes add to emissions, for example, by burning crop residues and felled trees, slash and burn agriculture, and forest fires. About 80 percent of forest clearing is to create more land for agriculture. Cattle graze on, and are fed by, most of this land. This is the leading cause of deforestation. It is estimated that meat production alone is responsible for 14 percent of global greenhouse gas emissions. Logging and mining also lead to deforestation but are much less a factor than clearing land for agriculture.

THE CARBON CYCLE

As noted in Figure 3.1, the amount of CO_2 in the earth's atmosphere has steadily increased from 300 ppm to 430 ppm in only about 175 years. This sharp rise is what distinguishes the current era from ancient times. It has been accompanied by a sudden increase in the earth's average temperature. "Temperature anomaly" refers to the *difference* in current temperature from the historical average temperature. The carbon rise is significant, because an increase of just 1.0 ppm in carbon dioxide concentration is equivalent to 7.86 billion metric tons **more** carbon dioxide in the atmosphere. It is hard to visualize such a large number. Let's ask Charlie.

Charlie: How many elephants would it take to equal 1.0 ppm more of CO_2?

Wow. Almost eight billion metric tons. That would be a lot of elephants! In the figure, each elephant represents ten million elephants, so these 15 elephants represent a total of 150 million elephants. The average male African elephant weighs about 6.75 metric tons (about 15,000 pounds), so 150 million would add up to 1 billion metric tons, and 1 part per million of carbon dioxide is almost 8 times that! And guess what? There's only about 150,000 male elephants left in Africa today!

The fast carbon cycle is the real-time flow of carbon through life forms on earth and in the oceans. This flow is seasonal. Less carbon dioxide is emitted during the growing season when plants absorb more carbon dioxide, and more is emitted during other times. It is a massive annual flow of carbon. Imbalances in the fast carbon cycle due to human-made emissions of carbon dioxide are the cause of today's global warming.

THE SIGNIFICANCE OF CARBON 14

Most of the carbon in atmosphere carbon dioxide is C-12, the carbon found in nature. A tiny percentage of atmospheric carbon is carbon-14 (C-14), the radioactive isotope of carbon. C-14 does not exist in nature. It is formed in the atmosphere when cosmic rays interact with nitrogen (80 percent of the atmosphere), converting nitrogen-14 (N-14) to C-14. As strange as it may seem, changes in C-14 concentration provide definitive proof that human activity is causing global warming. The explanation is complicated and we will not go into it here. [18]Suffice it to say that in 1955, a scientist named Hans Suess discovered that the atmospheric concentration of C-14 had decreased by 2.5 percent from the value in 1890.[19] Since the earlier date followed shortly after the beginning of the Industrial Revolution in the U.S., he postulated that what was happening was that the combustion of fossil fuels (coal, oil, and natural gas) was diluting the CO_2 in the atmosphere, so the ratio of C-12 to C-14 was *increasing*.

Oil, natural gas, and coal are called *fossil* fuels, because they are derived from the decayed remains of ancient plants and animals that died millions of years ago. They were formed so long ago that any C-14 has completely decayed. Fossil fuels only contain C-12. Suess saw the decrease in atmospheric C-14 as proof-positive that human activities were increasing the amount of atmospheric C-12, thereby diluting the amount of C-14 compared to its historical equilibrium value, a certain indicator that human activity is increasing greenhouse gases.[20]

Five countries and the European Union account for 63 percent of global CO_2 emissions. They are, in order of importance, China, U.S., India, the EU, Russia, and Brazil. In less than a decade, China's CO_2 emissions doubled! In 2024, China led the world in greenhouse gas emissions with 16 billion metric tons CO_{2eq}, 30 percent of the global total. China was followed by the United States with six billion metric tons, 11.3 percent of the global total. Table 3.3 lists the six top emitters.

Although China has the highest total emissions, the United States still has the greatest per capita emissions. See Table 3.3 for the comparison with other major emitters.

Country	Emissions (billion metric tons)	Percent (of global total)	Emissions (mtco$_2$eq/person)
China	16	30	11.1
USA	6	11.3	17.6
India	4.1	7.8	2.9
Eu 27	3.2	6.1	7.3
Russia	2.7	5.1	18.6
Brazil	1.3	2.5	6.0
Subtotal	33.3	62.8	N/A
Rest of World	19.7	37.2	1-3
Totals	53	100	7.26

Table 3.3: Top Greenhouse Gas Emitters, 2023. [21]

What about natural sources of carbon dioxide? Forest fires contribute to the carbon dioxide entering the atmosphere, but the amount is small compared to fossil fuel use. Likewise, volcanoes emit carbon dioxide, but only one-hundredth or less when compared on an annual basis to the amount due to burning fossil fuels.[22]

As China and India take steps to raise living standards for very large populations, their demand for energy will increase—more vehicles, air-conditioning, and increased agriculture production. The same is true for the balance of the lesser developed countries of Africa and Asia. This will be followed by increased greenhouse gas emissions until these countries can switch to renewable forms of energy.

In addition to the growing demand for energy due to population growth and lifestyle changes, the rapid growth of electric vehicle charging stations will also add to the growing load on electrical distribution systems. This would increase global warming if additional electricity is not generated by renewable sources. Likewise, experts are predicting a growth in electrical demand due to the new field of artificial intelligence (AI). Potentially, there could be a surge in electrical demand for very large computer installations that are being constructed. This is another potential source for increased greenhouse gas emissions.

In 1988, the United Nations established the Intergovernmental Panel on Climate Change (IPCC). In 2014, after detailed studies by a group of world climate experts, the IPCC issued a special report urging that emissions be reduced to "net zero" by 2050, to keep the earth's average temperature rise to less than two degrees Celsius (°C), and preferably less than 1.5°C above the preindustrial times average temperature.[23,24] "Net zero" means that any remaining human-caused emissions would be offset by withdrawals from the atmosphere.

In 2024, global human-caused carbon dioxide emissions set a new record projected to be 41.6 billion metric tons, up from 40.6 billion metric tons in 2023, a 2.5% increase, according to the Global Carbon Budget.[25] Carbon dioxide emissions make up about 75 percent of total greenhouse gas emissions, the remainder consisting of methane, nitrous oxide and fluorinated gases.

The 2°C limit is a best estimate of the maximum temperature rise beyond which the effects of global warming are unpredictable. [26]It is a judgment call. The current forecast is for the earth's average temperature increase to be about 2.5°C to 3.0°C (5.4°F) by the end of this century. [27]In 2025, after only a decade, it is clear that this goal will not be achieved.

4

DANGERS OF GLOBAL WARMING

As we discussed in Chapter 3, increases in greenhouse gas concentrations in the atmosphere lead to increases in the earth's temperature. This has an increasingly adverse effect on the earth's climate and population. The Intergovernmental Panel on Climate Change (IPCC) recommends that actions be taken to limit the earth's temperature increase to no more than 2.0°C, and preferably to no more than 1.5°C. To stay under this 1.5°C limit, we would have to reach net zero, no net human-caused greenhouse gas emissions, no later than 2050.

The 1.5°C temperature limit was exceeded in 2024, the warmest year ever recorded. This year was influenced by a strong El Niño that contributed to the observed temperature increase. It will take a few years of additional data to determine how much of the increase in 2024 was permanent and how much can be attributed to El Niño. Global emissions continue to grow so that the 1.5°C and 2.0°C limit will be exceeded at some point in the future.

The IPCC's 2.0°C limit is significant. It is not a sharp threshold beyond which something terrible will happen. It is considered to be the upper limit of present-day natural variability, especially in the tropics . Beyond 2.0°C, our ability to predict the effects of higher temperatures becomes increasingly uncertain.

CLIMATE CHANGE VERSUS WEATHER

It is important not to confuse climate change with the weather. Weather is the present or recently recorded measures of temperature, humidity, precipitation, and wind speed. The weather can change frequently and rapidly near-term and with the seasons. Weather is very variable and has occasional extreme or infrequent events such as a 100-year storm or flood. Climate is the average weather in a specific location over a longer period of time, like decades or more.

In discussing global warming, it is often difficult or impossible to differentiate between changes in the weather and climate change. Was a recent severe storm or flood due to climate change or is it a weather event where a similar event was recorded 100 years ago? It can be a combination of both—an extreme weather event that is more severe due to climate change. It may be impossible to differentiate between climate change and the weather.

A third factor may be human-caused. Storms or floods may have more severe impacts because of more structures being built on coastlines, flood plains, or other vulnerable areas.

THE IMPACT OF CLIMATE CHANGE

Some climate changes are more clearly attributable to global warming than others. Furthermore, climate change effects are very dependent upon the temperature increase—the higher the temperature, the greater the effect. Uncertainty about the effect increases with temperature. With a high degree of confidence, we can say that heat waves, sea level rise, and coastal floods are related to global warming. Other weather events are strongly influenced by global warming, such as droughts, hurricanes, forest fires and floods. But some events, such as tornadoes, don't seem to be linked to global warming.

Tipping points, the possible abrupt and irreversible changes that could occur as temperature increases are discussed. These events are difficult or impossible to predict and are not usually included in climate change forecasts. However, they are real possibilities that should not be ignored.

THE IMPACT OF TEMPERATURE ON CLIMATE CHANGE

The actual temperature increase due to global warming greatly impacts climate change. It is useful to think in terms of the following temperatures:

- 1.5°C, the IPCC's preferred limit for global warming.
- 2.0°C, the IPCC's upper limit for global warming beyond which climate effects are increasingly unpredictable.
- 3.0°C, approximately the upper limit of current forecasts of global warming by 2100, 75 years from now.

Table 4.1 summarizes the impact of these temperatures on climate change. Major actual and potential climate changes are discussed in the paragraphs that follow.

We are already experiencing irreversible climate changes, such as heat waves, longer wildfire seasons, sea level rise, loss of sea ice and permafrost melting in the Arctic, and melting of the Greenland and Antarctica ice caps.

TEMPERATURES RISE

The planet is experiencing rising temperatures in the land and the oceans. These temperature increases are causing the effects discussed in the following paragraphs.

HEAT WAVES

One of the most certain effects of global warming is an increase in the frequency and severity of heat waves. Heat waves affect people in a number of ways, including heat stress, heat stroke, and dehydration. The impact is severe for children, pregnant women, and the elderly. Labor productivity is reduced as temperatures increase, especially for outdoor and manual laborers. If productivity goes down, it affects incomes for communities relying on subsistence farming that requires hard manual labor.

In addition to heat waves, many regions are seeing higher average temperatures. For example, winters are warmer and spring comes earlier. There is more rain instead of snow. In some regions, growing seasons are longer, while others, mainly tropical areas, see higher temperatures reducing plant growth.

Possible Climate Effects
Temperature increase (1.5 ºC) Some irreversible changes, much of which can be accommodated through adaptation. • Severe heat waves • Longer wildfire season • Some crop yield reduction • Some freshwater shortages • Sea level rising • Flooding and coastal erosion • More severe hurricanes • Permafrost melting and sea ice retreating in the Arctic • Greenland and Antarctic ice caps melting

<table>
<tr><td>

Temperature increase (2.0 ºC)

Exceeding temperatures beyond which effects of warming are less predictable and risks triggering a tipping point.
- More frequent and much more severe heat waves
- Longer wildfire season
- Increased crop yield reduction
- Greater freshwater shortages
- Sea level rising faster
- More flooding and coastal erosion
- Severe hurricanes
- Permafrost melting, releasing carbon dioxide and methane
- Arctic increasingly ice-free
- Greenland and Antarctic ice caps melting faster

</td></tr>
<tr><td>

Temperature increase (3.0 ºC)

Can no longer predict full effects of global warming. One or more tipping points might be exceeded, possibly accelerating global warming with positive feedback effects.
- More frequent and much more severe heat waves
- Sea level rise of a meter or more
- Lose all coral reefs
- Melting Arctic permafrost releases carbon dioxide and methane adding to global warming
- Several regions unhabitable for parts of the year
- Crop failures
- Mass migration causing political problems and conflicts

</td></tr>
</table>

Table 4.1 Impact of Temperature on Climate Change

DROUGHTS

Global warming has led to droughts around the world. The effects are particularly noticeable in the western United States, Sub-Sahara

Africa, the Mediterranean and in Australia. In the western U.S., lack of rainfall has produced the driest 22-year period in at least the last 1,200 years. In addition to the lack of rainfall, hydrologists are concerned about the depletion of groundwater there.[28]

Droughts reduce crop yields and affect livestock production. The availability of fresh water decreases, and forests become more susceptible to wildfires and pests. These effects are especially felt in areas such as the Sahel, a large region crossing Africa south of the Sahara Desert. Droughts can force people to migrate in search of food and better living conditions.

SEA LEVEL RISE

Sea level rise is easily corelated with global warming. As discussed in Chapter 1, 90 percent of the heat from global warming is eventually absorbed by the ocean, the earth's largest heat sink. Sea level rise is about nine inches since 1880 and is increasing about 1.4 inches every ten years. This rate of increase is double that of the 20th century. It is dependent upon the rate of global warming. If greenhouse gas emissions are not reduced, sea level is projected to increase by an additional 1.0 foot by 2100 and could be higher if, for example, there is a rapid collapse of a major ice sheet in the Antarctic.[29]

Actual sea level rise varies by location due to uneven heating of the ocean, the uplifting or subsidence of land on the coast, and variations in wind patterns. Sea level rise increases the likelihood of coastal flooding due to storm surges. Around the globe, about 200 million people live within three feet of mean sea level.

HURRICANES BECOME STRONGER AND MORE INTENSE

Global warming will also pose a risk of hazardous weather.

Meteorologists know that there is a link between ocean warming and storm severity. Warm ocean water is the source of the energy that powers hurricanes and provides the moisture leading to rainfall and flooding due to these storms. Higher ocean temperatures increase the severity of hurricanes but not necessarily their frequency. Higher sea levels also increase the impact of storm surges from hurricanes.

Cyclones and typhoons are the same type of storm as a hurricane. In Southeast Asia, the Pacific Islands and east Asia, cyclones and typhoons are more destructive.

When Hurricane Harvey hit Texas in 2017, Houston's flood control system was supposed to handle a "100-year storm," defined as 13 inches of rain in 24 hours. The problem is that the "100-year storm" has already happened eight times in the last 27 years. Harvey was super-powered by warmer than usual Gulf waters. Besides massive rainfall, it caused a 15-foot-high storm surge along the coast. Within days, some parts of Houston had experienced 50 inches or more of rainfall—more than what is usual for a year. Tens of thousands of homes were flooded, and frantic residents were forced to evacuate.[30]

LONGER WILDFIRE SEASON

Global warming dries out vegetation and increases drought conditions that lengthen wildfire seasons and makes wildfires more frequent and intense.[31] These effects are most severe in the forests of the western U.S. and the boreal forests in North America and Russia. Climate change is the main cause of increased fire weather in the U.S., and these wildfires contribute to carbon dioxide emissions. Land use and poor forest management practices also contribute to wildfires.

Other areas of the world are also affected by more frequent and more severe wildfires such as the Mediterranean area: Italy, Greece and Turkey.

CHANGES IN PRECIPITATION PATTERNS

Global warming changes precipitation patterns around the world through increased evaporation in some areas and drier conditions in other areas. As the atmosphere heats up, it can hold more moisture. This can lead to more intense rainstorms that increase the risk of flooding.

The polar vortex or jet stream is affected by the more rapid heating of the Arctic.[32] The polar vortex becomes more unstable, affecting weather in the northern hemisphere where about 85 percent of the world's population lives.

Two of the earth's potential tipping points involve the monsoon rains in west Africa and India. Rainfall can become more concentrated in fewer but heavier rains that lead to crop damage and flooding. If the monsoons become unstable, it would affect the food supply for millions of people.

GROWING SEASONS LENGTHEN IN SOME REGIONS

Growing seasons are increasing in the temperate zones: North America, Europe, northern China and Russia. The average length of the growing season in the continental U.S. has increased by more than two weeks. This change has been accelerating since about 1970. [33]However, rising temperatures lead to hotter and drier conditions and lower agricultural productivity in other areas, especially the tropics. Many crops are stressed when temperatures exceed 90 to 95ºF.[34]

ARCTIC LOSES SEA ICE AND PERMAFROST MELTS

As stated earlier, the Arctic is heating up faster than the global average, causing loss of sea ice and accelerating permafrost melt. Permafrost is frozen ground in the Arctic region. Melting permafrost is a dangerous condition because it releases carbon dioxide and methane that have been trapped in the frozen plant material. It is estimated that in the Arctic, permafrost has about twice as much carbon trapped as is currently in the atmosphere. Permafrost makes up large areas of Alaska, Canada, Siberia, and the Tibetan Plateau. No one can say with certainty how much of it could be released as the Arctic warms.

ICE SHEETS IN GREENLAND AND ANTARCTICA MELT

The ice sheets in Greenland and Antarctica contain about 99 percent of the world's fresh water. This is enough fresh water to raise sea levels substantially as they melt. Since 2000, the stability of these ice sheets has deteriorated and the rate at which they are melting has increased, especially in the case of the Greenland ice sheet.

A potential tipping point is the breaking up of the Antarctica ice sheet extending into the ocean as global temperatures increase. Should this occur, the rate of sea level rise would increase. This ice sheet is a barrier limiting the flow of ice from the Antarctic ice cap into the ocean, preventing sea level rise. Melting of these ice sheets affects the Atlantic Meridional Overturning Circulation, a huge ocean current that takes warm water from the tropics to the Arctic. A reduction in the flow of this current would affect the climate globally.

DESTRUCTION OF CORAL REEFS

Coral reefs are especially vulnerable to climate change. Reefs are projected to lose between 70 percent and 90 percent of their area at 1.5°C of warming, and over 99 percent at 2°C. This will compromise food supplies, tourism, and protection of the coastline in some locations. [35]

The carbon dioxide levels in the ocean are rising at the same rate as the atmospheric levels, except that in deep water, CO_2 is accumulating more quickly. When carbon dioxide dissolves in salt water, carbonic acid is produced. Corals are living organisms and increased acidification has the potential to kill them, and they are a vital part of the food chain that supports the world's fisheries.

Increased CO_2 levels also cause lower aragonite (a form of calcium carbonate) saturation levels in the oceans around the world. This makes it more difficult for marine organisms to build shells and skeletons. [36]

Moreover, warm water is less dense and holds less oxygen than cold water. A dissolved oxygen content less than 2 mg of O_2 per liter is called "hypoxic" (hypoxia is inadequate oxygen in tissue) and is generally bad for fish and shellfish. Oxygen depletion is spreading with a potential serious effect on fisheries while diminishing a major source of food.

CROP FAILURES AND FAMINES [37]

Some of the effects listed above affect food supplies mainly due to lower crop yields resulting from higher temperatures, water scarcity, and more intense rainfall and flooding in some areas, especially in the most vulnerable regions of the world. Higher temperatures can also accelerate the activity of pests and diseases that affect crops. Increasing temperatures reduce the output of many crops including grains such as rice and wheat. Temperature effects are in addition to other potential damage caused by droughts or flooding.

INSECT INFESTATION

Warmer temperatures extend the time that insects are more active, thereby lengthening breeding seasons. Warmer temperatures also allow insects to migrate to areas that were previously too cold. Droughts also diminish the defenses of plants to resist pests.

In North America, the Pine Beetle has infected forests, killing trees. In East Africa, locusts have devastated crops. Warmer weather increases the spread of diseases carried by mosquitoes, fleas, ticks, and rodents.

SPECIES MIGRATION

Biologists know that every species has an optimum climate range for reproduction and survival. It depends not only on temperature, but also on rainfall and seasonal changes, including winter freezing.

Over one million species of plants and animals are facing extinction due to human activity. The causes are farming and land use practices that are destroying habitat, overfishing, global warming, land and water pollution, and the spread of invasive species, including bacteria, insects, and plants. [38]

DISEASES

Global warming is a major threat to human health around the world. Less developed countries are more vulnerable. Global warming compounds many of the problems faced by communities, such as inadequate housing, insufficient food supplies, water quality and availability, and exposure to toxins, diseases, and other health hazards. Increased stress can affect mental health as well as physical health.[39]

Many people are infected by tropical diseases that are water borne or hygiene related. Rising seas and flooding caused by more severe storms compromises drinking water, the disposal of human waste, and increases the risk of waterborne diseases caused by pathogens such as bacteria, viruses, and protozoa. Saltwater incursion can also contaminate water resources.

DESERTIFICATION

Deserts cover about one-third of the world's land area and are expanding. Part of this is due to overgrazing and other human-caused unsustainable land use practices.

Higher global temperatures that cause droughts and changes in rainfall patterns that kill off vegetation are a major cause of desertification. Some vulnerable areas such as the Sahel, a strip of land south of the Sahara, are especially vulnerable. The Sahel is home to about 400 million people that depend upon subsistence agriculture and livestock herding. This region is heating up about 1.5 times faster than the global average. The effects of higher temperatures accompanied by overgrazing and other problems are making this area less habitable.

MASS MIGRATIONS AND CONFLICTS

A number of factors attributed to global warming contribute to people choosing to migrate to other areas for food or a more hospitable climate. Factors include droughts and water shortages, crop failures, flooding, and storm damage. Social tensions can increase as people compete for food, fresh water, and grazing land. The United Nations estimates that over 300 million persons suffered forced migrations in 2024.

Global warming is a problem for those concerned about the threat from terrorism. Fewer young men will be able to support themselves as small farmers, so they will flock to already overcrowded cities looking for jobs that don't exist in the numbers required.

NATIONAL SECURITY IMPLICATIONS

Military bases and installations are negatively affected by extreme weather, flooding, and wildfires. In the U.S, some of the most vulnerable facilities are along the coast, such as the U.S. Navy base in Norfolk, Virginia. The Department of Defense has been surveying coastal facilities and making plans to raise them or relocate them to higher ground.

The military is asked to respond to natural disasters such as Hurricane Harvey. An increase in the number and severity of climate change disasters such as storms and flooding will put additional burdens on the military.

TIPPING POINTS

There is broad consensus that greater global warming increases the risk of unanticipated changes, some of which are potentially large, irreversible and self-perpetuating. Self-perpetuating means that the effect has positive feedback leading to greater changes in the future. An example could be the melting of permafrost in the Arctic that releases carbon dioxide and methane that in turn increases global warming, leading to greater melting of permafrost.

Tipping points involve the breakdown of some trend, a sudden and irreversible change such as the collapse of a bridge or other structure due to poor maintenance, overloading, or some other triggering event. They are the "straw that broke the camel's back."

- Shutdown of the Atlantic Meridional Overturning Circulation, the massive ocean current that transports heat from the tropics to the Arctic affecting the weather in Europe and other regions. Europe could become colder, and other regions hotter.

- West Antarctic ice sheet disintegration. A breakup of this ice sheet would lead to a substantial increase in sea level.

- Amazon rainforest dieback. This is the largest rainforest in the world, about twice the size of India. Ongoing destruction of this rainforest reduces the amount of CO_2 naturally absorbed by this forest.

- West African monsoon shift would decease rainfall in this already vulnerable area.

- Permafrost and methane hydrates. Permafrost sequesters huge amounts of carbon and methane that would release carbon dioxide and methane, adding to global warming.

- Coral reef die-off. Coral is very sensitive to ocean temperatures. These reefs support fish populations that are an important food source.

- Indian monsoon shift would affect rainfall during the monsoon season and effect India's food supply.

- Greenland ice sheet disintegration (melting) is accelerating and contributing to sea level rise.

- Boreal forest shift. These forests are in the cold regions in the northern hemisphere and account for 30 percent of the world's forests. They are huge carbon sinks. Like the Amazon rainforest, the destruction of these forests reduces the amount of CO_2 that is naturally absorbed.

At present, it is not possible to forecast any of these events with certainty, and they are not included in forecasts of global warming. These and other non-linear effects are unpredictable, but potentially devastating possibilities.

Some occurrences such as the depleting of sea ice in the Arctic and the melting of the Greenland ice sheet may have already passed a tipping point and are irreversible since the earth's temperature will only increase.

How much of air pollution is due to use of fossil fuels?

Air quality has improved considerably due to power plant scrubbers and cleaner burning vehicle engines. Many of us have become accustomed to the remaining air pollution and accept it as we do the weather, something to be tolerated. But air pollution remains as a major health problem, causing an estimated 100,000 deaths each year in the U.S. and a wide range of illnesses from asthma to some cancers. Air pollution aggravates respiratory infections and other health problems. Children and the elderly are especially vulnerable. Globally, an estimated 8 million die from air pollution each year, most of which can be attributed to fossil fuels. Several countries have much worse air pollution than the U.S.

Fossil fuels—a major source of greenhouse gas emission—are responsible for pollutants such as nitrous oxides (85%), sulfur dioxide (100%), PM2.5 (85%) and organic compounds (40-50%) that cause health problems. PM2.5 is particles that are small enough to be sucked into the lungs and is perhaps the most dangerous pollutant. The transition to renewable energy, electric vehicles, and heat pumps would eliminate most of this pollution.

5

FOSSIL FUELS

There are many different energy sources available today. Some have been known for thousands of years, while others are in early stages of development. This chapter discusses fossil fuels—historic energy sources—and discusses availability, economics, environmental effects, and impact on global warming and climate change. Without energy from fossil fuels, our modern living standards and the technologies available today would not be possible.

THE DAWN OF THE FOSSIL FUEL ERA

Prior to the Industrial Revolution in England in the late 1700s, power was provided by human and animal muscle with some assistance from water wheels and windmills. Heat was largely provided by wood for cooking, heating, metal working, and other purposes. A few locations had access to peat or coal. The Industrial Revolution can be said to have begun in 1712 in England when Thomas Newcomen invented a steam pump for removing water from mines. But the real impetus came 50 years later, when, in

1764, James Watt invented a greatly improved, more efficient steam engine. By 1850, the Industrial Revolution was underway with fixed steam engines powering factories and mobile steam engines powering ships and railroads. At this time, similar developments were taking place in the United States.

Edwin L. Drake drilled the first oil well in Pennsylvania on August 7, 1859. This development replaced the early uses of whale oil and initially was used for illumination and lubrication. Subsequently, Drake's discovery created an industry that expanded worldwide and delivered many benefits to humankind. The automobile created a huge demand for liquid fuels. In the 1950s, natural gas became a useful fuel, first for heating and power generation, then as a raw material for plastics and other materials.

In spite of all this progress, nearly three billion people still depend on the burning of biomass (wood, charcoal, crop residues, and dung) and coal in rudimentary stoves or open fires to meet basic needs for household energy.

Fossil fuels are abundant and relatively inexpensive. The world will not run out of fossil fuels for a very long time. To date, they have powered the industrialization of the global economy, using basically 100-year-old technology to provide most of the world's energy. See Table 5.1. Over the past 100 years or more, an extensive global infrastructure has developed to extract, process, transport, and make use of fossil fuels.

Environmental degradation resulted from the extraction and use of these fuels. Oil spills, strip mining for coal, and methane leaks from natural gas are examples. Fossil fuels are by far the largest source of the world's air pollution. Over time, these negative effects were partially mitigated. Much later, the world became aware of greenhouse gas emissions and their effect on the earth's temperature and climate. This has to change, and will change, but change will not come easily.

Total world energy consumption—driven primarily by fossil fuels—has been growing about 1.5 percent per year, compared to the growth rate for the global economy of about 3 percent per year. This indicates that world energy efficiency has been increasing about 1.5 percent per year, so a positive development. Overall, renewable energy, mainly solar and wind with utility battery storage, is growing rapidly, but still account for less than 20 percent of global energy use.

A little acknowledged fact is that the United Kingdom and the United States "own" a significant part of all the greenhouse gases currently in the atmosphere. As of 2023, global cumulative emissions were 24 percent for the U.S., 15 percent for China, 4 percent for the UK, and 3 percent for India, the leading contributors.[40]

When burning coal, oil, and natural gas to produce one million British thermal units (BTUs) of energy, different amounts of carbon dioxide are produced. Coal is the fossil fuel that produces the most carbon dioxide per unit of energy produced. The average for various types of coal is 212 pounds/MBtu; diesel oil is 163 pounds/MBtu; while natural gas produces the least, 117 pounds/MBtu. See Table 5.2 for a comparison of fossil fuels.

Energy Source	World Percent	United States Percent
Coal	26	8
Oil	32	38
Natural gas	23	36
Fossil fuels subtotal	81	82
Nuclear power	5	9
Hydro, solar, wind	14	9
Renewables subtotal	19	18
TOTAL	100	100

Table 5.1: U.S. and Global Energy Sources

THE GLOBAL ECONOMY IS POWERED BY FOSSIL FUELS

For more than 200 years or so, the world has depended upon power and heat from fossil fuels to provide the comfort and conveniences we enjoy today. The production and use of fossil fuels is a huge global business that employs millions. It will be very difficult to make a transition away from fossil fuel use voluntarily. A rapid transition is not practical. It is likely it will take a very long time to reach the IPCC's target of net zero emissions.

For the twenty-year period (1980-2000) following the Arab Oil Embargo and its sharp increase in oil prices, CO_2 emissions remained relatively flat due to global efforts to increase energy conservation and efficiency in response to the higher cost of oil. [41]If pre1970s oil prices had continued, emissions would be much greater today. After about 20 years of flat per capita energy use, per capita CO_2 emissions again increased in parallel with increasing GDP as oil prices declined due to supply increases. The long-term effects of the 1970s oil price hikes clearly show that higher energy prices lead to conservation and efficiency improvements

Reducing global energy consumption is an important measure for reducing carbon dioxide emissions. But it is unlikely that it can be effective as the primary means of combating global warming. Conservation and improving energy efficiency will help. However, the only effective way to combat global warming is to replace fossil fuels with renewable forms as the world's primary source of energy.

In other words, the best way to keep fossil fuels in the ground is to replace them with renewable energy that is cheaper. It must also be as accessible and reliable as today's fossil fuels. Solar, wind, and hydro energy have to become the main sources of energy, augmented by next-generation nuclear if radioactive waste management, siting, and other nuclear issues can be resolved. Hydroelectric power is another important resource in the limited areas where large dams can be constructed. Some fossil fuel displacement can come from conservation and efficiency improvements. It is unlikely that all fossil fuel use can be eliminated. Some will be required as a raw material in manufacturing and for some transportation modes.

Fuel Type	Advantages	Potential Problems
Coal	The most abundant and lowest cost fossil fuel in many countries. Can be transported globally by rail and ship.	Produces more CO_2 per unit of energy produced than other fossil fuels. Is a major source of air pollution.
Oil	Abundant global supply. Global transportation network for low-cost distribution. Source of all liquid fuels: gasoline, diesel fuel, aircraft fuel. Is a feedstock for plastics and other products.	Produces CO_2 and other greenhouse gases. A major source of urban air pollution. No readily available substitutes for liquid fuels. Oil price increases quickly impact global economy. Price can vary considerably.
Natural Gas methane	Fossil fuel with lowest CO_2 emissions. Low levels of other pollutants. A feedstock for fertilizer and plastics production.	Produces CO_2. Methane is a greenhouse gas; leakage contributes to global warming. Requires pipelines for economical transportation. Price can vary considerably.

Table 5.2 Fossil Fuel Comparisons

COAL

Coal has been an inexpensive fuel and is abundant in many but not all countries. For example, Japan and Korea do not have significant coal deposits. Coal is still an important fossil fuel used for the production of electricity on a global basis. Using coal in China and India to meet a growing demand for electricity is one of the reasons carbon dioxide emissions have been increasing in these countries. Table 5.3 shows global coal consumption by major users. China and India are increasing their use, while the U.S. (and the EU) are reducing theirs.

Year	2020	2021	2022	2023	2024
World	7.5	8.0	8.0	8.0	8.0
China	4.0	4.3	4.3	4.3	4.3
India	0.9	1.0	1.1	1.2	1.2
Other Asia	1.0	0.9	0.9	0.9	0.9
U.S.	0.4	0.5	0.5	0.4	0.4
Eu	0.4	0.4	0.5	0.5	0.4
Rest of World	0.7	0.8	0.8	0.8	0.8

Table 5.3 Global Coal Consumption (billion metric tons) [42]

Coal is carbon. (C is the chemical symbol for carbon). When coal is burned to produce energy, the energy is derived from converting carbon and oxygen into carbon dioxide. Coal includes other contaminants such as sulfur. During combustion, sulfur is converted into oxides of sulfur, a health hazard. Burning coal also produces particulates that are discharged to the atmosphere. There is a wide variety of coals available depending upon where it is mined. Some types have more impurities than others, generating more pollution as well as carbon dioxide.

To reduce carbon dioxide emissions, reducing the use of coal should be the first priority where it is possible. But some countries currently do not have good options to reduce coal without technical and financial support from the international community. Worldwide there are over 2,400 coal-fired power stations, generating about one-third of the world's electricity. While China and India are building dozens of new coal-fired plants, other countries, such as Germany and the U.S., are closing theirs. In April 2019, renewable sources in the U.S. generated more electricity than coal for the first time.[43] The largest coal-fired plants in the U.S. are being closed as a result of no longer being economical.

OIL AND ITS DERIVATIVES

Oil and natural gas are hydrocarbons. So, their molecules contain hydrogen atoms as well as carbon atoms. When hydrocarbons are burned to produce energy, a large percentage of the energy produced is from combining hydrogen and oxygen to form water (H_2O). This water leaves the combustion process as water vapor. Oil is the source of essentially all liquid fuels such as gasoline, diesel oil, and aviation fuel. Oil is also the cheapest fuel to transport globally. There is a huge global infrastructure for oil transportation using oceangoing tankers and pipelines.

At one time, there was a concern that the world was running out of inexpensive sources of oil and that oil prices would increase dramatically. However, new technologies for the exploration and extraction of oil have led to a global surplus of oil *and lower oil prices.* Fracking (oil and gas produced by injecting water under high pressure into subterranean rock formations to create or open fissures to extract oil or gas) in the U.S. has had a major impact on the cost of producing oil and natural gas. It has also made it possible to tap oil deposits that were considered to be depleted or uneconomical to exploit. This led to a surge in U.S. oil and gas production so that the U.S. became the top producer, followed by Saudi Arabia as the second largest and Russia in third place. U.S. oil production was 20.4 million barrels/day (Mbbls/day) in 2024.

Oil prices are variable depending upon global supply and demand and events from recessions to military conflicts. From

2020 to 2025 the price had been oscillating from $U.S. 50 to $U.S. 60 per barrel, to as high as $U.S. 124 per barrel in 2022 as a consequence of Russia's invasion of Ukraine.

In 2024, the U.S. was the world's single largest consumer of oil at 20.3 Mbbls/day, followed by China at 16.4 million bbls/day. (See Table 5.4). The U.S. has been energy independent since 2019, producing more oil, natural gas and coal than it consumes. The U.S. continues to import some oil and refined oil products for various reasons but these imports are exceeded by oil exports. The U.S. also exports natural gas as LNG and coal.

NATURAL GAS

Natural gas, essentially methane ($CH4$), is presently the cleanest fossil fuel, producing the lowest amount of carbon dioxide for a given energy output.[45] The U.S. currently has an abundance of natural gas due in part to fracking. Until recently, natural gas was inexpensive compared to other fossil fuels, ranging from $U.S. 2.00 MBtu to $U.S. 3.00 MBtu pipeline cost (early 2020), or $7-$10 per MBtu for commercial and residential customers. Prices increased when Russia invaded Ukraine. In Europe, gas prices until recently had been twice U.S prices but shot up to about six times U.S. prices in 2022 after the Russian invasion of Ukraine and the sabotage of the Nord Stream 1 and 2 pipelines. In 2025, natural gas prices ranged from about 4$/MBtu in the U.S. to three times higher (12$/MBtu) in the EU.

Because of the low cost of natural gas in the U.S., it has been displacing coal in the U.S. for electricity generation and is the major contributor to the decline in greenhouse gas emissions. However, methane is a potent greenhouse gas, and the leakage of methane during the extraction and transportation of natural gas adds to greenhouse gas concentrations in the atmosphere. The demand for natural gas is increasing as a fuel and as a feedstock for chemicals, plastics, and fertilizer production. Natural gas is also the main fuel for home heating. So, increased demand could eventually raise the cost of natural gas in the U.S. and other countries.

Year	2020	2021	2022	2023	2024
World	91.19	97.08	99.57	100.2	103.2
U.S.	17.2	18.8	18.9	19.0	20.3
China	14.4	14.9	15.0	16.6	16.4
India	04.7	4.8	5.2	5.5	5.6
Saudi Arabia	03.4	3.6	3.9	3.9	4.0
Russia	03.3	3.5	3.6	3.8	3.8
Rest of World	48.9	1.5	53.0	51.4	53.4

Table 5.4: Global, U.S., China, India, Saudi Arabia, and Russia Oil Consumption Mbbls/day [44]

The cheapest way to transport natural gas is by pipeline. Pipelines are expensive to build and require a right of way (property easement), which is often contested by property owners. The availability and capacity of pipelines can limit access to natural gas. Transporting natural gas across the ocean, for example from the U.S. to Japan, is very expensive. Natural gas has to be converted to a liquid at a very low temperature and transported in expensive, special-purpose ships. At the receiving end, it has to be converted back into a gas. This is energy-intensive and requires a heavy capital investment. This limits the potential for exporting natural gas compared to oil and even coal. Table 5.2 compares fossil fuel alternatives.

SUBSIDIES

Producers of fossil fuels are heavily subsidized in most countries. In 2023, global subsidies of fossil fuels exceeded U.S. $620 billion. This figure is significantly lower than the record high of over $1.2 trillion in 2022, but it still represents a substantial amount of support for fossil fuels. It includes tax breaks as well as cash subsidies in some countries for gasoline and electricity. The biggest subsidy by far is the ability to discharge greenhouse gases and other pollutants into the atmosphere essentially for free. With these costs, the total subsidy is more like $7 trillion. The cost and damage related to these emissions, mainly global warming and air pollution, are paid for by the general public, not by the emission generators. See Table 5.5.

Subsidy (2022)	Amount ($trillion/yr)	Percent (of total subsidies)
Underpricing fuels	1.0	15
Air pollution, global warming, other environ-mental effects	6.0	85
TOTALS	7.0	100

Table 5.5: Estimated Fossil Fuel Subsidies

Subsidies promote the use of fossil fuels and delay the transition to renewable sources of energy by making fossil fuels appear less expensive than they are. Remove the subsidies, and fossil fuels would be more expensive and the transition to renewable wind and solar energy sources would be accelerated. Subsidies come in various forms, including direct government payments, tax breaks, and underpricing of fossil fuels according to the International Energy Agency (IEA).[45] The IEA data shows that subsidies are broadly split between electricity, oil, and natural gas, with coal receiving minimal support according to a Reuters report.[46]

CARBON FEE OR TAX

The most significant and effective thing that the U.S. could do to demonstrate that it is serious about fighting global warming and setting an example for others, is to implement a national fee on carbon. A fee or tax on carbon has been proposed as one way to "even the playing field" for renewable energy to compete with fossil fuels. Charging a fee for carbon emissions is part of getting the incentives right. As the cost of carbon emissions increases, renewable alternatives become more attractive to energy users without the need for subsidies. So far, the U.S. federal government has not taken this necessary action.

Globally, a number of countries have implemented some form of carbon fee. A carbon fee would also need a border adjustment tax or similar mechanism to penalize imports from countries that don't have a carbon tax. A global agreement on a carbon tax, complete with a border adjustment, is improbable.

Applying a reasonable cost to the discharge of carbon dioxide and other pollutants into the atmosphere would be a major incentive for greater use of conservation and renewable sources of energy. A carbon fee would also add to government revenues, as opposed to being a government expense for grants, subsidies, and tax breaks.

About 40 countries have proposed or enacted legislation that prices carbon. The total is divided roughly 50/50 between countries with Cap-and-Trade and those with a carbon fee. Carbon fees range from $1/metric ton of carbon (mt) in Mexico to a high of $139/mt in Sweden, with the majority between $10/mt and $25/mt.

Any carbon fee or surcharge would have to be applied gradually to allow economies and individuals to adjust to a higher cost of fossil fuels.

The estimated impact a $50 per ton carbon fee would have on the retail prices of commonly used fuels is shown in Table 5.6. One benchmark to price carbon dioxide emissions would be to charge emissions the same price as the cost of removing carbon dioxide from the atmosphere. This could be hundreds of dollars per metric ton.

Allocating the cost of subsidies (Table 5.5) to global CO_2 emissions results in an estimated cost of about $190/ton of carbon discharged into the atmosphere. Excluding the 85 percent social costs results in a cost of about $29/ton of CO_2 emitted.[47] Imposing either would increase the cost of fossil fuels to politically unacceptable levels.

Fuel Type	Average Price	Carbon Fee $50/ton		Carbon Fee $100/ton	
Gasoline	$2.17 gal	$0.44	20%	$0.89	40%
Natural gas	$2.31 MMBtu	$2.31	115%	$5.30	230%
Heating oil	$2.10 gal	$0.51	24%	$1.02	48%

Table 5.6: Impact of a Carbon Fee on Fuel Costs

A lower, more acceptable tax would be helpful, with gradual adjustments over time.

FOSSIL FUEL DOWNSIDES

There are a number of disadvantages associated with using fossil fuels.

Air pollution remains a major health problem, causing an estimated 100,000 deaths each year in the U.S. It contributes to a wide range of illnesses from asthma to heart disease and some cancers. Fossil fuels, the major source of greenhouse gas emissions, are responsible for most air pollution.

Another downside is oil spills from oil tankers, pipelines and other sources. Natural gas leaks discharge a powerful greenhouse gas, methane, into the atmosphere. Leaks come from gas wells, pipelines, leaks from power plants to furnaces and other sources. Flared gas, the burning of natural gas that is not or cannot be captured in oil fields and refineries, also adds to pollution and greenhouse gas emissions.

Oil and gas drilling produces waste products, especially from fracking that uses large amounts of water and chemicals that has to be disposed of. Some of these waste products can contaminate ground water and pollute the land.

Land use for oil and gas fields has to be abandoned eventually as oil, gas and coal are depleted. This land is no longer useable for other purposes such as agriculture or recreation. It is not practical to fully restore this land for other uses. In addition, who pays for the cleanup?

For wind and solar farms, the land can be used indefinitely for solar and wind electricity production even if solar panels and wind turbines have to be replaced at the end of their useful lives. These locations will not run out of sunlight and wind.

OUTLOOK FOR FOSSIL FUELS

Coal-fired power plants no longer competitive with low-cost natural gas generation are being retired in some countries. In 2025, scheduled closure of U.S. fossil power plants is 12,300 MW total. Of that total, 8,100 MW is coal, 2,600 MW is natural gas, and 1,600 MW is oil-fired plants.[48] The U.S. utility industry had plans to close fully half of the U.S. coal-fired generating capacity by 2026, but the current administration is offering grants to extend some plants' operations. Likewise, coal is being phased out in the EU, while China and India are increasing the use of coal for electricity generation and manufacturing.

The development of fossil fuels, beginning with coal, then adding oil and natural gas to power industry and generate electricity, created an industry that expanded worldwide and delivered many benefits to humankind. Later, it became apparent that this industry also had a dark side—aspects that caused widespread human illnesses and death through air pollution and through global warming.

The future demand for fossil fuels might rise because of an increase in electricity demand caused by more electric vehicle charging stations and powering large AI data centers. This would primarily impact coal, nuclear, and natural gas to the extent that renewables are unable to meet the demand. Fossil fuel demand may decrease, depending on the extent to which renewable energy sources become more competitive (cheaper than natural gas), or due to advances in technology, or because electric vehicles replace internal combustion engines in greater volumes than expected.

There is no likelihood the world will run out of fossil fuels anytime soon. Industry estimates of economically recoverable resources and years to depletion at current rates of use are shown in Table 5.7.

Fossil Fuel	Economic Recoverable Resource	Supply Lifetime (Years)
Coal	1.6 trillion short tons	Over 100
Crude oil	1.5-1.7 trillion barrels	Over 50
Natural Gas	7,000 trillion cubic feet	Over 50

Table 5.7: Economically recoverable fossil fuels and years to depletion at current rates of use.[49]

HOW MUCH OF THE ENERGY IN FOSSIL FUELS IS ACTUALLY USED TO PRODUCE POWER AND HEAT?

Much of the primary energy, about 32 percent of the energy content of fuels, is lost in production, processing and transportation. Still more—about 45 percent—is lost as waste heat due to the inefficiency of power plants, engines, and furnaces that produce useable power and heat. This fact is often overlooked in the debate concerning fossil fuels and renewables.

In total, on average, only about 33 percent of the primary energy in fossil fuels is useable. What this tells us is that to replace fossil fuels with renewables, we do not have to replace all the primary energy, only that portion that winds up as useable power and heat. Electricity is more efficient to transport and use than fossil fuels. Doubling electricity production using solar and wind could replace most fossil fuel uses, avoiding the need to mine, drill wells, and refine oil and gas. Most fossil fuel uses could be replaced, except for a small part necessary for some manufacturing and transportation processes. Sources:

1. https://leahy.substack.com/p/673-of-all-energy-generated-in-the
2. https://rmi.org/the-incredible-inefficiency-of-the-fossil-energy-system/

The cheapest resources have been largely exploited in the last 100 years. If the cost of renewable energy sources continues to decline, fossil fuels will face price competition in a market where their production costs are rising. It is conceivable that the world may experience a slow decline in the demand for fossil fuels while not running out of them.

6

RENEWABLE ENERGY ALTERNATIVES

Energy users have over 150 years invested in producing and using fossil fuels. Any transition to alternatives that do not emit greenhouse gases will be difficult. We should not underestimate the time, expense and effort required. Fortunately, there are practical alternatives to the fossil fuels we are using today. Some alternatives are already being adopted and are growing rapidly, such as solar energy. Others are under development and may or may not prove to be useful, such as green hydrogen as a means of storing and transporting energy.

DEFINITION OF RENEWABLE ENERGY

From a global warming perspective, an energy alternative must produce useable energy for power and heat without discharging greenhouse gases. It must also be cost-competitive to avoid unacceptable energy price increases. It must be abundant to replace all the fossil fuel energy we use while supporting future increases in the energy the world needs to support population growth and rising living standards. There are other practical considerations, such as the ease which any energy alternative can be produced, transported, and stored. All energy sources will have certain advantages and disadvantages that need to be overcome or accommodated.

CAPACITY COMPARISONS: RENEWABLES VERSUS FOSSIL FUELS

Comparing various power sources for electricity production can often be confusing. The most common measure is power output measured in megawatts. However, this measurement needs some qualifications. Not all power sources are designed to produce their rated power output continuously. Renewable energy, solar and wind, are subject to the time of day and the weather that affects them, like solar radiation and wind speed.

Some power sources such as a nuclear power plant can produce close to their rated output, such as 1,200 MW continuously except for refueling about every 18 to 24 months or so. A refueling can take 30 days or more. However, nuclear power plants cannot increase or decrease their power output quickly to accommodate changes in electricity demand. Rather, they are best suited for base load power generation in conjunction with other electricity sources that can rapidly adjust output to respond to increases and decreases in electricity demand.

Capacity factor is a measure of, on average, how much peak power is produced over time. See Table 6.1 for a summary. Solar farms in good locations operate at or near their peak capacity for about eight hours in the middle of the day when the sun is shining. Battery storage is needed to store electricity to meet demand when the sun isn't shining. Wind farms in good locations provide steady output but operate below rated capacity for much of the day based upon wind speed. They also need battery storage to meet electricity demand when demand exceeds supply.

Power Plant Type	Typical Capacity (Megawatts)	Capacity Factor (Percent)	Comments
Gas Combined Cycle	400-800	40-75	Largest source of electricity in U.S.
Gas Turbine (peaking plant)	50-300	5-15	Can be cycled up or down quickly to match demand
Nuclear power plant	1,100-1,700	90-95	Steady power output
Solar farm	300-1,500+	25-30 in optimum locations	Operates during sunlit hours
Wind farm onshore	500-2,500+	35-40	Requires steady winds
Wind farm offshore	500-2,500+	40-50	Requires steady winds
Battery storage	100-1,600+	Typically, 4-12 hours duration	Stores electricity from solar and wind for use during peak demand and helps stabilize the grid

Table 6.1 Capacity Factors for Various Electricity Generating Plants

Batteries have rated capacities the same as a power plant. This is the maximum amount of power that can be supplied by the battery. Batteries are also rated by the number of hours they can supply peak power, typically four to twelve hours. A battery rated for 100 MW and four-hour capacity could, for example, provide 50 MW (half its capacity) for eight hours.

TODAY'S SITUATION

The use of renewables for electricity production is growing rapidly in the U.S. and globally. In the U.S. in 2024, electricity capacity additions were about 62.8 gigawatts (GW). According to the U.S. Energy Information Agency (EIA) about 63 percent was for solar, 23 percent was for battery storage and 13 percent was for wind, compared to 4 percent for natural gas. This emphasis on renewables is projected to continue.

Globally, in 2024, about 93 percent of electricity capacity additions were for renewables, a growth rate of about 15 percent per year. Of the 585 GW additions, about 77 percent was for solar and 32 percent were for wind. In addition, 160 GWh of utility-scale battery storage was added in 2024, a 68 percent increase over the prior year.

The first step in the energy transition is to produce all electricity using renewables—mainly solar and wind. Nuclear power, hydro, and geothermal will help. Wherever possible, current fossil fuel used for transportation, or to produce power and heat, must be converted to electricity, for example, using electric vehicles (EVs) and replacing gas furnaces with heat pumps or microwave or induction furnaces.

RENEWABLE ENERGY ALTERNATIVES.

Table 6.2 compares advantages and problems associated with various renewable alternatives.

Energy Source	Advantages	Potential Problems
Solar	Fuel-free, available in many locations. Cost of solar-produced electricity is competitive and decreasing. Electricity easily distributed via the electric grid. Can be deployed rapidly to meet electricity demand.	Not universally available. Intermittent output and varies by time of day, time of year, and with the weather. Requires large land area. Not practical at certain latitudes.
Wind	Fuel-free, available in many locations. Cost of wind-produced electricity is competitive and decreasing. Electricity easily distributed via the electric grid. Can be deployed rapidly to meet electricity demand.	Not universally available. Intermittent output and varies by time of day and with the weather. Requires large land area; hazard to birds.
Nuclear	No greenhouse gas emissions. Steady electricity production at full power, not subject to variations in weather and sunlight.	Most expensive power plants to build. Cannot be rapidly deployed to meet growing electricity demand. Depleted nuclear fuel has to be processed and safely stored forever. Public concerns about nuclear safety and waste disposal led to decline of nuclear power in U.S. Has to be cost-competitive with solar and wind with battery storage.

Hydro	Dependable source of electricity.	Limited growth potential. Requires dams and large reservoirs.
Geo-Thermal	Uses heat from the earth's core. Steady electricity production at full power, not subject to variations in weather and sunlight.	Expensive plants to build. Limited to specific geographic areas. Has some growth potential. Has some pollution such as from impurities in steam from the earth.
Biofuels	Can be substituted for gasoline and diesel fuel.	Releases CO_2 when consumed. Assumes CO_2 removed from the atmosphere as plants regrow. Requires large land area to grow feed stocks. Uses fossil fuel to power farm vehicles and to produce fertilizer. Competes with food crops for land.

Table 6.2: Advantages and Potential Problems of Renewable Energy Sources

SOLAR ENERGY

Electricity from solar cells is the largest and fastest growing form of renewable energy by far. It has unlimited growth potential. Solar panels produce about 25 percent of their rated power averaged over the day. To provide steady power throughout the day, battery storage is needed to store electricity for use when demand exceeds supply. Solar cells convert a portion of the sun's radiation directly into electricity. There is no fuel and no moving parts.

THE "NEW SAUDI ARABIAS" FOR SOLAR.

Globally, the potential for solar and wind is more than adequate to provide all the energy the world requires. Top solar regions globally, the *"New Saudi Arabias"* will be substituting solar energy for fossil fuels. High-voltage transmission lines or undersea cables will be used instead of pipelines or oil tankers to transmit energy as electricity to other countries that lack adequate renewable resources. These solar- and wind- intensive locations can also produce fuels such as hydrogen or ammonia or synthetic fossil fuels for export. Their low-cost energy could be used locally for the production of important materials such as aluminum or for energy-intensive industries and even data centers.

This is already happening. Saudi Arabia, the Middle East's largest oil producer, is also one of the world's fastest growing markets for solar power. Saudi Arabia plans to have half its electricity come from renewables by 2030. The Saudis are working to use renewables to produce green hydrogen for export and exploring ways to export electricity from renewables to Europe.[50]

Some possible sites for major solar power installations are located around the globe. For example:

- Chile's Atacama Desert
- North Africa (Sahara)
- The Middle East
- Southwestern United States & Northern Mexico
- Australia (Outback, Northern Territory, Queensland)
- China (especially Inner Mongolia deserts)
- India (Thar Desert and Rajasthan)
- Brazil (Northeast region)
- Southern Europe (Spain, Italy, Greece)
- Vietnam & Southeast Asia.

WIND

Large wind turbine technology has advanced and continues to improve in cost and performance. Larger and more powerful wind turbines are being developed. They can be almost 900 feet tall with a rotor diameter of about 400 feet and a maximum power output of about 12 MW. Wind power is second only to solar as the renewable energy source with the most growth potential.

Wind turbines are fuel-free and do not have greenhouse gas emissions. However, they are large rotating structures that require maintenance. They can also be damaged in a severe storm. The typical design life for a wind turbine is 25 years.

The output of wind turbines depends on the average wind speed over a twenty-four-hour period. A typical wind turbine will produce about 40 percent of its rated power over time.

There are two basic types of wind turbines—onshore and offshore. Those in the ocean can be on a foundation on the sea bottom in shallow water. In deep water, the wind turbine must be on a floating platform that is anchored to the sea bottom. Undersea cables transport electricity from the wind turbines to the shore. Offshore wind turbines have the advantage of steadier and stronger winds over the ocean, but they are more expensive to install, maintain and operate. The public has opposed wind turbines in scenic areas.

NUCLEAR

Nuclear power is an abundant source of carbon-free energy and is available 24/7. All reactors, including new designs, will have depleted nuclear fuel that must be disposed of, then the radioactive waste must be stored essentially forever. Technical leadership for the design, construction, and operation of nuclear power plants has moved to China. U.S. capabilities have declined.

The U.S. nuclear power industry was booming in the 1970s and 1980s. In 1990, there were 112 reactors operating in the U.S. and output peaked at 20 percent of electricity production. The industry has been essentially dead in the U.S. for over 30 years. To revive nuclear power, the problems that led to the decline of nuclear need to be resolved:

- Nuclear power plants are complex and the most expensive power plants to build. Several plants experienced major cost overruns and schedule delays. Some projects were abandoned. This led the utility industry to question the economics of nuclear power.

- Public concerns about nuclear safety, especially following the Three Mile Island nuclear accident in 1979, led to longer licensing delays and expensive new safety requirements that added to costs and delays.

- Siting became a problem. Most people do not want to live near a nuclear power plant, even if the experts say that new designs are much safer. This raises the question of where new nuclear power plants will be located.

- The spent fuel problem must be resolved. This is a political, not a technical problem that other countries have solved. In over 50 years, the U.S. government has not been able to reach a politically acceptable solution for the storage of spent nuclear fuel. Today, all spent fuel is stored at the reactor sites, even for nuclear plants that have been shut down.

- Nuclear power has to be competitive with other sources of electricity based on the cost per kilowatt hour, especially in deregulated electricity markets. They benefit from their ability to deliver near full power 24/7, contributing to the stability of the electrical grid. This could command some price premium.

Today, the main cost competition is from power plants fueled by natural gas. In the future, the main competition will come from renewable energy, solar and wind, with large utility-scale battery storage needed to match supply and demand to stabilize the electric grid.

- The U.S. no longer mines or produces the enriched uranium to fuel nuclear power plants. Imports accounted for 99 percent of the uranium oxide used to produce nuclear fuel in 2023.[51] This material is imported from Canada, Australia, Russia, and other countries. Imports from Russia were banned in 2024, but waivers have been given for imports from Russia through 2028. Efforts are being made to revive uranium production in the U.S.[52]

The U.S. is the world's largest producer of nuclear electricity, producing about 18 percent of its electricity. There are 93 nuclear reactors in 54 power plants in 28 states. The average age is about 45 years. Many are reaching the end of their service lives and will be shut down in the foreseeable future. Only one new nuclear power plant has been completed in the U.S. in the past 30 years, and no large new nuclear plants are being planned.

New reactor designs are under development in the U.S. and other countries. There are three main types: small modular reactors (SMRs), molten salt-cooled reactors, and sodium-cooled reactors. It will take time to get regulatory approval, fund, build, and test these reactor designs and to determine their cost, performance, and safety. As a footnote from two old nuclear engineers, these are not new concepts, despite what the developers claim. They all have been tried before, in the 1960s and 1970s, and abandoned in favor of pressurized and boiling water-cooled reactors.

HYDROELECTRICITY

There is limited potential for adding to hydroelectric power generation. There are not many remaining dam sites to be developed, and new dam projects are opposed by environmental groups. They object to damming rivers and flooding large areas needed to create reservoirs. Also, as global warming progresses, there may be droughts that could affect stream flows and reduce hydro potential.

GEOTHERMAL

Geothermal uses heat from the earth's core. It requires drilling into hot geological formations and injecting water that is converted to steam that drives a turbine on the surface. Geothermal can produce a steady and reliable source of electricity at full power and is not affected by the weather and time of day as is solar and wind. However, corrosion caused by salts and other materials that can contaminate the steam coming to the earth's surface can be a problem and may lead to some air pollution.

The U.S. presently has the largest geothermal power production, with about 4,000 megawatts (MW) of installed capacity. Most of this capacity is in the western states and Hawaii where the underground heat source is nearer the surface, with California being the biggest user of geothermal energy.

Geothermal power plants today are typically small with an average size of about 40 MW. The largest geothermal complex today is the Geysers geothermal complex in California. It covers about 45 square miles and includes 22 geothermal power plants producing about 725 MW of electricity.

The potential for geothermal increases with the ability to drill deeper wells to reach hot spots in the earth's crust. Investments are being made to improve and expand the use of geothermal energy. This potential is limited to locations with favorable geological conditions.

ENERGY STORAGE

Energy storage, mainly batteries, are needed to store electricity for small mobile applications such as cell phones, electric vehicles, and large batteries to store variable energy from solar and wind farms. Large utility-scale systems are fixed installations with very large capacities. The use of utility-scale batteries is increasing rapidly as more wind and solar energy is added to the electric grid. Mobile applications from vehicles to laptops require batteries that are lightweight and compact. Electricity storage for non-mobile applications can be much larger and heavier to provide more storage capacity.

Battery technology today is recognized as the main problem limiting the range of electric vehicles. Batteries are also the most expensive component of most electric vehicles and largely determine the price of the vehicle. Major investments are being made to improve battery technology so that the range is increased and the charging time is reduced.

Because of its light weight, the preferred means of storing electricity for mobile applications is the lithium-ion battery. It is used in everything from laptops to automobiles. Lithium is the lightest metal in the periodic table and has the greatest electrochemical potential for electricity storage by weight. It is the preferred choice for mobile applications today. The increasing production of lithium-ion batteries for vehicles and other electricity storage applications is reducing the cost of these batteries and expanding their use.

Lithium-ion batteries are also used in large, utility-scale, energy storage systems for electric utilities. Other battery designs such as flow batteries are being developed for large, utility-scale energy storage. Since size and weight aren't a limitation for non-mobile batteries, other storage technologies might be superior to lithium-ion batteries for fixed installations.

HYDROGEN

Hydrogen can be a green fuel in that its combustion yields water (H_2O), not CO_2, so long as it is produced by electrolysis using electricity from renewable sources. Hydrogen-powered fuel cells are more efficient than gasoline or diesel combustion engines.

However, hydrogen is not a source of energy but an energy carrier like electricity. Hydrogen is very reactive, and there is no free hydrogen—no hydrogen mines or wells. There are several ways to produce hydrogen for fuel—by electrolysis, which separates water into hydrogen and oxygen using electricity, or by natural gas reforming, which produces hydrogen from natural gas. This process uses fossil fuels to produce hydrogen and is not green energy.

If solar and wind power can produce abundant, low-cost electricity, it may be possible to use low-cost renewable electricity to produce hydrogen at a price that would make it a potential fuel source for fuel cells, gas turbines, and internal combustion engines.

Hydrogen is a gas at room temperature. So, it must be compressed at high pressures to reduce its volume before it can be used as a vehicle fuel.

FUEL CELLS

Fuel cells allow the direct conversion of a fuel such as hydrogen into electricity without combustion. Hydrogen-powered fuel cells do not emit greenhouse gases. If hydrogen is the fuel, the only products are electricity, water, and heat. Because there isn't any combustion, fuel cells do not produce pollutants such as sulfur dioxide or nitrous oxides, a greenhouse gas.

Fuel cells can exceed 60 percent efficiency, compared to a maximum of about 40 percent for internal combustion engines and most power plants. They can use a wide range of fuels and feedstocks and can provide power for systems as large as a utility power station and as small as a laptop computer.

Prototypes of fuel cell-powered vehicles are on the road today, such as Toyota's Miria. Some very large fuel cell power plants are under development. A 79 MW fuel cell power plant is operating in Korea and can supply power to 250,000 homes. The world's largest fuel cell power plant will be a 110 MW plant that is under development, also in Korea.

THE TEXAS EXPERIMENT

A very important experiment is being conducted in Texas. It will demonstrate the merits of electricity production using fossil fuels versus solar and wind, based upon cost, availability and reliability.

We know that Texas is a fossil fuel powerhouse accounting for about 42 percent of oil production and 29 percent of our natural gas production in the U.S. But it is not as well known that Texas also leads the U.S. in electricity production using renewables. The Lone Star State also leads in the installation of new solar and wind capacity and battery storage.

HOW IS THIS POSSIBLE? THERE ARE SEVERAL REASONS:

- Texas has a state-wide electric grid managed by ERCOT (The Electric Reliability Council of Texas). Thanks to this grid, it is easy to connect the best areas for solar and wind farms to electricity customers throughout the state.

- Texas also has a competitive electricity market. Producers have incentives to use the lowest cost sources for electricity while passing the lower costs on to customers. Low electricity prices also result from the use of low-cost natural gas, not just renewables. The average electricity rate in Texas is 10 cents/

kWh compared to the national average of 13 cents/kWh. Compare this to the average rate in California of 26 cents/kWh.

- Texas favors new development and it is easier and quicker to get new energy projects approved and implemented than most other states.

- The state leads in the addition of utility scale batteries to store electricity to smooth out supply and demand to stabilize the grid. ERCOT actively manages the grid in real time. Large electricity users can be required to adjust their electricity use when total demand is high. Additions to solar, wind, and battery capacities have added to electricity reserves and strengthened the state's ability to reduce or avoid blackouts and brownouts during times of peak demand.

Texas's economy will get a boost from its inexpensive, plentiful electricity. Low-cost electricity from renewables will also help the state become a leader in the production of green hydrogen, a fuel for the future.

California is a close second to Texas in renewable energy. But California has among the highest electricity rates in the continental U.S. and imports about 25 percent of its electricity. This is in spite of having excellent potential for solar and wind projects. California does not have a competitive retail electricity market. The Public Utilities Commission approves electricity rates requested by the state's three large utilities. This high and rising cost of electricity in California works against the greater use of electricity from renewables for transportation, home heating, and other uses being mandated by the state.

CHINA EMBRACES RENEWABLES

China leads the world in the manufacture and use of solar and wind energy, battery storage, electric vehicles, high-speed rail, and nuclear power. Since 2015, China has tripled its renewable energy investments and is investing more than any other country in 2025, over $600 billion.

In 2024, China accounted for 64 percent of new global renewable energy capacity additions. It installed 277 GW of solar, a record that is likely to be exceeded in 2025 and has also installed about 80 GW of wind. China's total capacity of 1,400 GW of renewable energy was completed about six years ahead of its target for 2030. Green energy is meeting about 80 percent of the increase in China's electricity demand.

China's goal is to electrify and decarbonize its energy system by 2060, when its energy will come from non-fossil sources. China's carbon emissions are set to peak by 2030 and decline thereafter. By 2060, China plans to get almost all of its energy from renewables; solar and wind (47%), nuclear power (19%) and hydro and biomass (20%). Fossil fuels would account for the remainder (14%). Some use will be made of carbon capture and storage where carbon dioxide emissions cannot be eliminated. Some synthetic fuels will be produced using carbon dioxide captured from the atmosphere or from exhausts.

WHY?

China has several reasons to aggressively pursue renewable energy and to electrify its economy.

China became the world's largest oil importer in 2013. Unlike the U.S., China does not have abundant oil or natural gas reserves. Renewables reduce the need for oil and gas imports and the foreign exchange needed to pay for them. Oil and gas imports are also a security risk being subject to a blockade during a crisis or conflict.

China also sees renewable energy and electric vehicles as major global growth businesses and aims to become the world's leading supplier.

China has serious air pollution, as well. The greater use of fuel-free renewables and nuclear power results in major reductions in air pollution.

In addition, China sees electricity as the best way to power its modern manufacturing economy and support new industries such as AI that require large amounts of electricity. This will increase China's global competitive advantage.

A SMART, NATIONAL ELECTRIC GRID

To support renewable energy, China has undertaken a massive upgrade of its grid. This includes the world's largest high-voltage transmission system to connect areas well suited for solar and wind farms to cities and industries in other locations. This is a smart grid that is designed to accommodate the variable nature of solar and wind. The backbone of this grid is ultra-high-voltage transmission lines to enable efficient transport of electricity over long distances. Battery storage and pumped hydro are used to smooth out supply and demand and stabilize the grid that includes a growing share of variable solar and wind electricity sources. Some major power users can be required to reduce electricity demand during supply shortages. Time-of-use pricing of electricity with higher prices during periods of peak demand shifts some demand to off-peak times.

SOLAR AND WIND FARMS

About 85 percent of the world's solar cells are manufactured in China. China's Tengger Desert Solar Park is the largest solar plant in the world. Four of the five largest manufacturers of wind turbines are Chinese companies. The largest wind farm in the world is being built in China—The Gansu Wind Farm—with a planned capacity of 20,000 MW. Today, China produces about 40 percent of its electricity using renewables, compared to Europe (also at 40 percent) and the U.S. (24 percent).

NUCLEAR POWER

Nuclear power is a smaller but important renewable energy source. China has 48 nuclear power reactors in operation, 12 under construction, and more about to start construction. China's policy is to have a closed nuclear fuel cycle under which it will recycle spent reactor fuel to reduce nuclear waste products and recover useable nuclear material to recycle into new reactor fuel.

ELECTRIC VEHICLES

China now leads the world in total vehicle production and is producing about 70 percent of the world's electric vehicles. Chinese brands are increasing market share in China over foreign brands manufactured there. To promote electric vehicles, China has about 13 million EV charging stations, about one station for every 2.7 electric vehicles.[53] In 2025, electric vehicles were slightly over half the light vehicles sold in China compared to the EU (30 percent) and the U.S. (9.6 percent).

BATTERIES AND RAW MATERIALS

Thanks to its lead in manufacturing mobile electronic devices, electric vehicles, and large utility-scale batteries, China is now leading the world in battery technology and in the production of the raw materials needed for batteries.

HIGH-SPEED RAIL

China's high-speed rail network includes about 30, 000 miles of track connecting cities across China. This rail network was started in 2008, the same year that California voters approved a bond issue for the initial funding of California's High-Speed Rail project. Now more than 70 percent of the world's high-speed rail track is in China. Almost all (97 percent) of major cities with populations of 500,000 or more are connected to high-speed rail. Europe has about 6,000 miles of high-speed rail and Japan has about 1,800 miles. The only high-speed rail project in the U.S., California's High-Speed Rail project, has completed only 70 miles of the planned 170-mile link in the past 15 years.

China is slightly larger (+16 percent) than the continental U.S. The U.S. has also constructed massive national transportation networks, but some time ago, and without the use of computers, modern communications, and today's construction equipment. Between 1860 and 1900, the U.S. added 170,000 miles of railroad track to complete a national network. The first transcontinental rail link, 1,900 miles long, was completed in 1869. By 1900, there were six transcontinental rail links.

Starting with congressional authorization in 1956, the U.S. built the world's largest highway network, 48,000 miles of limited access freeways largely completed in the 1970s to serve almost all urban areas with a population of 50,000 or more.

OUR RENEWABLE ENERGY FUTURE?

As discussed in this chapter, there are practical alternatives to fossil fuels. All energy sources, fossil fuels and renewables, have advantages and disadvantages. There are large renewable energy projects in the U.S., Europe, China, and India that demonstrate that solar and wind with an appropriate level of battery backup are increasingly viable and economical compared to fossil fuels.

These renewable sources will continue to benefit from ongoing technology improvements and increasing economies of scale.

Could the Sahara Desert provide all the world's electricity? What about electricity equivalent to all the world's energy?

The Sahara desert is 3.6 million square miles in area (9.2 million square kilometers). The average solar irradiance in the Sahara is approximately 2,000 kwh/m2/year.

A modern solar panel is about two square meters in size and thus could collect about (2 m2 x 0.22 x 2,000 kwh/m2/year) or about 880 kWh/year. Global energy use in 2025 is estimated to be 200,000 billion kWh/year. Dividing this by 880 kWh/panel, we find 23 billion panels, each two square meters, would be required. Thus, the panels alone would take about 4.9 percent of the Sahara desert. To find the actual total area required, it is necessary to make allowances for access roads, spaces between rows of panel, substations, transmission lines, and so on. Assuming this excess area is about equivalent to the panel area itself, about 10 percent of the Sahara desert would power the world.

To provide all the world's electricity (30,000 billion kWh/year, using the same assumptions   would require (30,000/200,000) x 10 percent = 1.5 percent of the Sahara Desert.

I'm guessing it would be lot cheaper than drilling for oil.

CONCLUSION

We are not thinking big enough about the use of renewables. Electricity is a clean, efficient, and easily transported form of energy compared to coal, oil, and natural gas. So, why can't Europe get a significant percentage of its electricity from solar farms in the Sahara? Australia and Singapore are planning a 1,700-mile undersea HVDC (high-voltage direct current) cable to transport renewable electricity from Australia to Singapore.[54]

Humankind is drilling for oil in the deep ocean and in the sands of the remote Middle East. We transport this oil thousands of miles in oil tankers that are the largest ships on the sea. Natural gas is shipped around the world as a super-cooled liquid in specially designed ships to areas that cannot be reached by pipelines. In addition, oil needs huge refineries to convert crude oil into useful products such as gasoline. None of this infrastructure will be required for electricity from renewables.

7

WHY IS IT SO HARD TO STOP GLOBAL WARMING?

The transition from fossil fuels to renewable energy is difficult. Energy users have invested over 150 years in producing and using fossil fuels. The time, expense, and effort required should not be underestimated. Chapter 9: "What it Takes to Stop Global Warming" describes a plan to overcome some of the major difficulties. Fortunately, there are practical alternatives to the fossil fuels in use today. We authors believe that most or all of the technology needed to deal with global warming is currently available. As we have described, some alternatives are already being adopted and are growing rapidly, such as solar energy.

GLOBAL POPULATION AND ECONOMIC GROWTH ARE DRIVING GLOBAL WARMING

There are two fundamental forces driving global warming. The first is the world's large and increasing population—more people, more energy consumption. Population growth is demanding more and more of the earth's resources. The second force is rising living standards throughout the world. As people's living standards improve, they use more energy. Since 2000, the global population has increased from 6.1 billion to 8.1 billion, a 33 percent increase. This is an annual growth rate of about 1.1 percent per year. Although the world's population is increasing, the rate of increase is slowing, expected to be 0.9 to 1.0 percent per year in the coming years. India surpassed China as the most populous nation in 2023. Half or more of population growth by 2050 will be in just eight Asian and African countries. The United Nations forecasts that the world's population will reach about 9.7 billion by 2050 and over 10 billion by 2100.[55] This increase in population will drive the demand for energy higher.

From 2000 to 2024, global living standards increased from $10,600 per capita to $17,900 (Adjusted constant 2017 dollars) or about 70 percent, or 2.2 percent per year. With higher living standards, people and organizations use more energy for vehicles, home appliances, increased lighting, and air-conditioning, for example. Efforts to date have not reduced greenhouse gas emissions. During this same period, greenhouse gas emissions increased from 37 to 57 $GtCO_2e$, a 60 percent increase or about 1.9 percent per year.

POTENTIAL PROBLEMS ARE NOT OFFSET BY OBVIOUS BENEFITS

The U.S. and the world have gone through transitions that were as disruptive or more disruptive than this transition from fossil fuels to renewables. These changes involved huge investments, large construction projects, disruptive changes in employment, and other upheavals. Some examples are the Industrial Revolution, the steamship, railroads, the automobile, electricity, globalization of the economy and the electronics revolution that brought us computers, the internet, cell phones, automation and artificial intelligence.

However, these disruptive changes were accompanied by major benefits for many if not most people and society so that, in general, these changes were supported by a majority of those affected. Giving up fossil fuels and making other changes to stop greenhouse gas emissions does not have similar obvious benefits. Most air pollution is eliminated, but transitioning from fossil fuels to renewables and the other necessary changes to stop global warming does not have any immediate or perceivable benefits for the average person.

INEFFECTIVE AND INCONSISTENT U.S. EFFORTS

The U.S. federal government has been inconsistent in efforts to stop global warming. President Trump withdrew the U.S. from the Paris Agreement for the second time in 2025. The current administration's main-concern is that efforts to slow or stop global warming are a threat to economic growth in the U.S. Global warming is not a major threat and should be ignored in favor of the greater use of the U.S.' abundant fossil fuel reserves. This administration is not content to eliminate support for greenhouse gas reductions and the participation in international organizations dealing with this problem. Efforts are being made to undermine those concerned about global warming and disband government research capabilities.

Previous actions to reduce greenhouse gas emissions have favored regulations, mandates, subsidies, and tax breaks and have not been very effective. There are mandates that cannot be met such as the federal government's target of net zero, no human-caused greenhouse gas emissions no later than 2050. California's goal of banning the sale of vehicles with internal combustion engines by 2035 is another example. These goals are unrealistic if they are not backed up with implementation plans, funding, and organizations with the authority and responsibility for oversight and implementation.

In 2022, President Biden signed into law the Inflation Reduction Act (IRA). The Inflation Reduction Act is an 840-page bill addressing inflation and health care costs as well as global warming. This bill is the biggest climate spending package in U.S. history, approving an estimated $360 billion for climate change over ten years mainly through grants, loans, subsidies, and tax breaks. But an inadequate effort was undertaken to establish means to select the best projects and audit spending and to determine what programs are effective and if objectives are being met. Much of this legislation is being rolled back by the Trump administration.

THE NEED FOR UNPRECEDENTED, PERHAPS UNACHIEVABLE, GLOBAL COOPERATION

Perhaps the biggest obstacle to dealing with global warming is the need for international cooperation, especially between China, the U.S. and India, the three largest emitters. A coordinated international effort is required with firm commitments for reductions. If only a few countries take action, others get a free ride and not enough will be done to solve the problem. The level of cooperation needed, and the long-term coordinated effort that will be required, is similar to that of the Allies in WWII. In the past, international cooperation has succeeded in solving serious global problems, for example in dealing with the hole in the ozone layer. So far, it has not worked for global warming.

There are effective international organizations in place which is a major step forward. The Paris Agreement is an international agreement signed by 195 countries specifically to deal with global warming. The Intergovernmental Panel on Climate Change (IPCC) is doing a good job in organizing available data and the latest research for policymakers and others. The IPCC and other international organizations issue comprehensive progress reports measuring our success toward stopping global warming. The

United Nations Conference of Parties (COP) is an annual meeting of countries that signed the Paris Agreement to review progress and future actions. Nationally Determined Contributions (NDC) are non-binding pledges stating each country's goals for greenhouse gas reductions. However, it is still up to each country to establish their objectives to reduce emissions and enact plans to achieve these objectives. The U.S. and other major emitters must set an example that other countries can follow for solving this problem.

THE U.S. MAY HAVE LESS INCENTIVE TO SWITCH TO RENEWABLES

The U.S. is the only large, developed country that is currently self-sufficient in oil and gas, thanks to the fracking revolution and its abundant fossil fuel reserves. Others—such as China, Japan, Europe, and rapidly developing India—have more incentive to embrace renewables to reduce their dependence on imported fossil fuels. China and India also have huge air pollution problems that can be reduced by the transition to renewables.

There is a danger that the U.S. will resist the growth of renewables to preserve its fossil fuel industry as long as possible and to protect domestic vehicle manufacturers. If so, the U.S. could fall further behind other countries in the development and production of solar panels, wind turbines, batteries, electric vehicles and nuclear power plants.

GLOBAL INEQUITIES IN ENERGY USE

Lesser developed nations in Africa, Asia, and the Middle East need to be allowed to increase their per capita energy use to provide more electricity for home and industrial use. They will need to substitute natural gas for firewood and to provide gasoline for more motor vehicles or acquire electric vehicles.

The level of domestic and international cooperation to effectively solve this problem and fairly share the associated costs and sacrifices may be beyond the current capabilities of today's domestic and international government institutions. Breakthroughs in international understanding and cooperation will be required.

MISINFORMATION

The first two chapters of this book summarize the irrefutable evidence that global warming is real, is increasing and is irreversible. In spite of this evidence, a number of people and organizations deny the facts and claim that global warming is not real or is something we cannot do anything about or worse, that we can live with the consequences associated with higher temperatures and climate changes.

Most people don't realize that global warming is a real problem that needs solving. Global warming is something they have heard about, and they may be concerned about rising temperatures and other climate changes. However, they are not personally affected very much, if at all. Much of the public discounts the seriousness of global warming. If it is real, they postulate, it will affect people far in the future or somewhere else.

Globally and in the U.S., the fossil fuel industry is a very big business that employs millions of people and is a major source of income, new investment, and profits and dividends for individual investors, investment funds, and pension funds. There are huge, vested interests that will resist changes that negatively impact this industry. This includes the vast automobile industry that has for over 100 years invested in designing and producing vehicles with internal combustion engines powered by gasoline and diesel fuel. Some people and organizations, especially those in fossil fuel industries, have deliberately published false information about the impact of global warming and what needs to be done about it.

Reading this book will hopefully allow you to identify and rebut the false arguments against efforts to stop global warming.

If you think that global warming is a hoax, not that bad, or something we cannot do anything about, then you are likely to object to the effort, changes, and investments required to stop it. Others do not want to admit that fossil fuel use has unacceptable side effects and want us to continue to use fossil fuels.

A common criticism is to point out the cost or difficulty of doing something to stop global warming without also mentioning the negative effects if the earth's temperature is allowed to increase which will affect our climate. Some common deniers' arguments are:

- Using renewable energy retards economic growth while greater use of fossil fuels is good for the economy.

- Global warming is overstated. There are better, more cost-effective ways to deal with climate change.

- There are other more important problems that we should be concerned about such as healthcare and poverty.

- Increasing carbon dioxide concentrations improves plant growth leading to higher agricultural yields.

- Global warming is a slow, linear process. We have time to adjust to climate change effects without stopping global warming. Exceeding a tipping point is very unlikely and should be ignored.

- Technology will bail us out if climate change becomes a real problem such as geoengineering, fusion power or yet to be discovered solutions.

- The cost to stop human-caused greenhouse gas emissions is more than we can afford.

CLIMATE CHANGE DENIERS

If you believe global warming is a hoax, not that bad, or something we can't do anything about, then you are likely to object to the investment and disruption that would be needed to eliminate fossil fuels and make other changes. You are a denier. You might argue that you're not a scientist, but you've heard about some scientists who deny that global warming is real. Some believe that global warming as a problem has been exaggerated and a crash effort to address it is not warranted and will fail. Global warming is overstated. There are better, more cost-effective ways to deal with climate change. A crash effort to get to net zero by 2050 to keep global warming under 1.5°C is impractical and wasteful. It won't succeed and will waste trillions of dollars.

Global warming should not be our biggest concern. There are other major problems facing the global population that are more important. Time and effort are better spent on health care, for example. Some maintain that the impact of global warming on rising living standards is fairly small. Global economic growth and rising living standards will continue to improve. There has been considerable progress for over 50 years, and this progress will continue. The world is getting better. The impact of global warming is linear and will be slow enough that the world can adapt to many of the effects of global warming such as coastal flooding at an acceptable cost. If needed, future technical solutions such as geoengineering are a backup plan.

So far, the effects of global warming have been linear. However, if global warming doubles over the next 50 years, that may not be true. Waiting will make the problem worse and harder to solve. Adaptation and prevention have to be a big part of any plan to reduce global warming impacts. However, prevention should take priority over adaptation as much as possible. There could be positive feedbacks that accelerate warming or some tipping points could be exceeded that lead to abrupt and irreversible changes such as massive ice melts that trigger accelerating sea level rise. We believe that waiting until the eleventh hour and then resorting to some unproven technology such as geoengineering or carbon capture is a high-risk experiment that may have unforeseen and irreversible consequences that we can ill afford.

PUBLIC EDUCATION

Educating the public and getting widespread acceptance that global warming is a serious problem and something must be done about it, is challenging but essential. There are several factors that work against getting public support. People have a tendency to believe what they want to be true. They don't want to be reminded of a growing global threat. There is a general belief that catastrophic events are rare and unlikely to affect them. It is human nature for people to want to "tune out" big problems.

There is a lack of comprehensive, readable, accessible information about global warming and what should be done about it. The public needs to know the facts, consequences, and alternatives available, to know that there are practical things that can be done. Much of the information available is complicated and written for scientists. People tend to rely on limited sources of information: the people they trust, newspapers, TV, or the internet. Overall, the press has done a poor job of educating the public. The issue is further confused by the amount of disinformation being disseminated by those associated with the fossil fuel industry to confuse the public and promote the idea that global warming is not an urgent problem.

In a recent Pew Research report, 64 percent of Americans say that climate change is currently affecting their local communities. Those living near the coasts are more likely to say that climate change is affecting their community. Almost half of those surveyed say that climate change is not an easy subject to understand. In spite of this public awareness, the Pew Research Center's survey of public opinion of the important issues facing the country lists strengthening the economy as top issue, as in recent past years. Climate change has ranged from number 18 out of 19 issues in 2019, to 17/21 in 2023 and to 18 of 20 in 2024. Most people worry about something else. See Figure 7.1.[56]

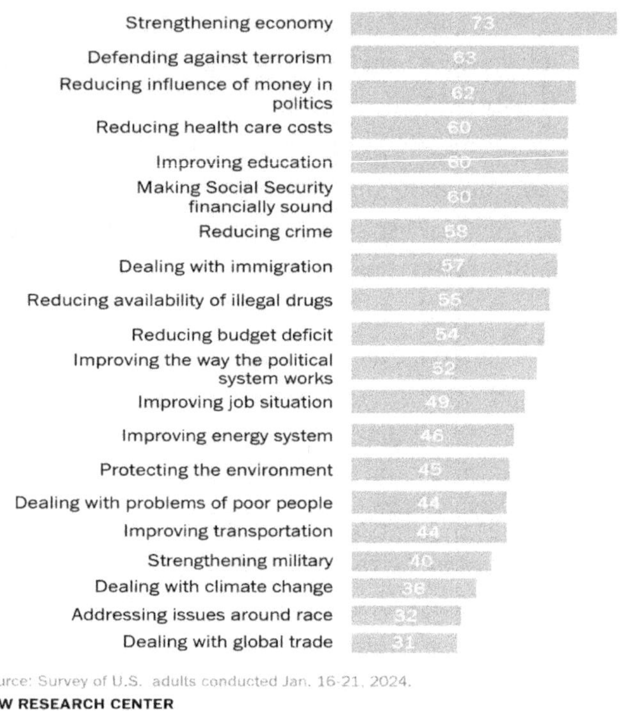

Figure 7.1 U.S. Public Policy Priorities 2024

Actions today such as curtailing the continuing use of fossil fuels may take 25, 50, or more years to have a visible effect. People may have to make significant sacrifices today to solve a problem that may not affect them very much. The problem is intergenerational. Steps need to be taken now to solve somebody else's problem in the future. This requires a long-term viewpoint that is too much to expect from populations that are struggling just to make ends meet, or those who are so focused on day-to-day survival that they cannot pay attention to long-term health needs, do not save anything for retirement, etc. These people are being asked to sacrifice today to improve the lives of future generations.

The world is faced with a difficult problem. The earth's temperature is about 1.5°C above the preindustrial baseline of 1850 to 1900. CO_2 levels have increased from 280 to 430 ppm. If aggressive actions to reduce the use of fossil fuels began immediately, CO_2 levels would reach about 500 ppm before human-caused emissions are reduced to a level that is equal to or below the level that can be absorbed naturally by the earth. The degree of cooperation, foresight, and sacrifice needed may be beyond current capabilities on a national or global basis. At the current rate of action, a severe crisis could occur first, before triggering effective actions. We will see a substantial increase in the earth's temperature with serious effects on the earth's climate. It's understandable why this problem is so difficult to solve. However, an effort must be made.

8

INTERNATIONAL EFFORTS TO STOP GLOBAL WARMING

The year 1957, the same year that Sputnik was launched, was designated the International Geophysical Year and led to more funding and a more coordinated international approach to climate research.[57] The International Global Atmospheric Research Program was established in 1967 to gather data for weather forecasting and climate research. That same year, it was calculated that doubling CO_2 concentration in the atmosphere would increase atmospheric temperatures by about 2°C compared to the preindustrial age baseline.

The U.S. National Oceanic and Atmospheric Administration (NOAA) was established in 1970 and became the world's leading source of funding for climate research. By the late 1970s, scientific opinions tended to conclude that global warming was a major climate risk. The U.S. National Academy of Sciences reported that it is credible that doubling CO_2 concentration in the atmosphere would increase the earth's temperature from between 1.5 to 4.5°C. The World Climate Research Program was also launched to coordinate international climate research.

1981 was reported as the warmest year on record at that time. In 1987, the Montreal Protocol of the Vienna Convention imposed international restrictions on the emission of ozone-destroying gases. This was to address the "hole in the ozone layer" problem. This is perhaps the first example of international cooperation to solve a global environmental problem. The following year, in 1988, the Toronto conference called for limits on greenhouse gas emissions.

The U.N.'s Intergovernmental Panel on Climate Change (IPCC) was established by the United Nations Environment Program (UNEP) and the World Meteorological Organization (WMO) in 1988. The IPCC was endorsed by the UN General Assembly in 1988.

The IPCC does not conduct original research. Its purpose is to collect the best scientific data and analysis and make this information available to policymakers and the public through reports written by experts in various disciplines.

HISTORY OF IPCC GLOBAL WARMING OBJECTIVES[58]

What is the origin of the IPCC's objective stated in the Paris Climate Agreement of limiting global warming to less than 2°C with a further objective to limit warming to 1.5°C? The 2°C limit was initially proposed by economist Richard Nordhaus in 1975 based upon his intuition that "more than 2°C above preindustrial levels would take climate outside the range of observations that have been made over the last several hundred thousand years."

The 2°C threshold was a judgment call and a reasonable objective at the time. It was high enough to be achievable in a specific amount of time. It was low enough to reduce the risks for unpredictable or runaway climate change. It also established a clear goal for international efforts to stop global warming.

In a 2018 special report, the IPCC recommended that efforts begin to limit global warming to no more than 1.5°C to protect people more vulnerable to climate change, to reduce the risk of reaching climate tipping points, and to further reduce the impact of global warming.

In 1996, the European Council of environmental ministers was the first major political organization to formally endorse this limit, stating that global average temperatures should not exceed 2°C above preindustrial levels.

OTHER EFFORTS

In 1997, 193 countries signed the world's first agreement on greenhouse gas emissions—the Kyoto Protocol. No temperature limit was specified at the time. The U.S. did not ratify the Kyoto Protocol.[59] Starting in about 2008, the Alliance of Small Island States (AOSIS) called for more ambitious limits and supported a limit well below 1.5°C to reduce the disastrous impact of rising sea levels.

In 2009, the Copenhagen Accord accepted the 2.0°C temperature limit as the central goal but said it would consider limiting the temperature rise to 1.5°C.

Finally, in 2015, the landmark Paris Agreement was signed under which 195 countries supported the long-term goal to limit global warming to well below 2°C and to pursue a goal of 1.5°C.

THE INTERGOVERNMENTAL PANEL ON CLIMATE CHANGE (IPCC)

The first IPCC report was issued in 1990. It concluded that global warming was occurring, and future warming was likely. The Fifth IPCC report was issued in 2014. This report confirmed that greenhouse gas emissions from human-caused sources are the main cause of global warming. This report also highlighted the risks associated with global warming and associated climate change. In addition, the report stated that the means are available now to limit global warming. This report established a target of trying to stabilize the earth's temperature increase to no more than 2.0°C above preindustrial levels.

THE PARIS AGREEMENT [60]

In 2015, the Paris Agreement was signed whereby all nations including the U.S. agreed to set targets for their greenhouse gas emissions and to report their progress periodically. This was a breakthrough international agreement addressing climate change. The main goal of the Paris Agreement was to limit the global average temperature rise above preindustrial levels to be less than 2.0°C, and preferably under 1.5°C. That same year, 2015, was the hottest year on record due in part to a strong El Nino reaching 1.0°C (1.8°F) above the preindustrial level. The level of CO_2 in the atmosphere exceeded 400 ppm for the first time in millions of years.

Under the Paris Agreement, each country was required to submit an action plan defining how they will achieve their self-determined emissions targets, their Nationally Determined Contributions (NDCs). There are no enforcement mechanisms and no penalties for countries that do not meet the emissions targets in their NDCs.

U.N. CONFERENCES OF PARTIES (COP)

The COP is an annual meeting to review the national commitments and emission inventories submitted by the Parties to the Paris Agreement. Based on this information, the COP assesses the effects of the measures taken by Parties and the progress made in achieving the ultimate objective of the Convention. The first COP meeting was held in Berlin, Germany in March 1995. The 2025 meeting, COP30, will be held from November 10 through 21, 2025 in Brazil.

EMISSION GAP REPORTS

The United Nations Environmental Program (UNEP) issues an annual Emissions Gap Report summarizing the results of current environmental studies and estimates of future greenhouse gas emissions. This estimate is compared to the goals stated in the Paris Agreement, the "Nationally Determined Contributions" (NDCs) stipulated by each party to the agreement. The purpose is to determine the gap between current emissions pathways and what is required to meet the established goals.

Annual Emission Gap Reports are essentially annual progress reports on the world's efforts to reach net zero by 2050. The latest report, *Emissions Gap Report 2024*, was issued in October 2024. [61]This report states that nations must have a massive decrease in their Nationally Determined Contributions (NDCs), if the IPCC goal of keeping global warming under 1.5ºC is to be met. The 2.0 degree C limit is also at risk.

Based upon current NDCs, the forecast is for a global temperature rise of 2.6-2.8°C by 2100. In addition, actual efforts to achieve current NDCs compared to pledges results in a temperature rise of 3.1°C by 2100.

9

WHAT IT TAKES TO STOP GLOBAL WARMING

It is obvious that the current administration is committed to rapidly expanding fossil fuel production and use in the U.S. This administration is eliminating incentives for renewable energy wherever possible and taking steps to deny that global warming is a problem that needs to be addressed. This is a major deterrent to making progress in the U.S. toward renewable energy and other actions needed to stop global warming.

Any progress that will be made will be based upon economics that favor renewable energy, incentives offered by some states, and the actions of individuals and organizations that choose to do something favorable.

Texas could be a positive example of what might be achieved today. Texas leads the U.S. in renewable energy production.

So, in spite of the administration's efforts to prevent actions to address global warming, what do we have to do? The needed actions will require significant changes in energy supply and use along with global cooperation among the largest countries emitting greenhouse gases: China, the U.S., Europe and India. All of this will take a long time. The list of necessary actions looks overwhelming. However, encouraging progress is being made today.

Even if we can't see today how best to close the gap to get to net zero, we have to start reducing greenhouse gases immediately, and as fast as possible consistent with avoiding recessions and other problems. For the present, we describe what we consider a realistic Action Plan to start reducing greenhouse gas emissions, knowing that achieving net zero by 2050 is impossible, given the difficulties that exist.

A REALISTIC FORECAST

A good starting point would be to make a realistic forecast of what can actually be done while avoiding a crash effort that cannot be achieved such as net zero by 2050. Attempting a schedule that cannot be met discourages supporters and gives climate deniers reason to discredit efforts to stop global warming. A realistic forecast is not available but for discussion purposes, we should anticipate that a successful effort would take about 50 years or more.

SUMMARY OF ACTIONS REQUIRED

- Upgrade and modernize the nation's electrical transmission and distribution networks to distribute up to three times today's electricity output.

- Improve energy efficiency and conservation to reduce future energy needs. Get more GDP with less energy. Improving efficiency by 1.0 percent per year reduces the amount of energy we use by about 18 percent in about 25 years.

- Produce all electricity using renewables, primarily solar and wind. Nuclear power and other green alternatives can help. However, solar and possibly wind farms are the technology that can be deployed fast enough to accomplish this.

- Replace fossil fuels with electricity wherever possible. Use electricity instead of fossil fuels to produce power and heat for industry, commerce, and residential needs.

- Transition transportation to electric vehicles. Transportation is the largest user of oil-based fossil fuels by far.

- Reduce emissions from methane, nitrous oxides, and fluorinated gases. Methane leakage could increase with the growing use of natural gas and LNG.

- Improve land use planning. Solar and wind farms should be placed in locations with the most sunlight and wind potential. Optimal locations can be developed without using scarce farmland or areas unacceptable to the public. Most states do not have the potential to produce most of the renewable energy they need. They will have to import electricity similar to today's need to import the fossil fuels they use.

- Develop fossil fuel alternatives for uses that cannot be converted to electricity such as airline travel and ocean shipping. There will be a need for synthetic fuels made from recycled carbon dioxide and electricity from renewables.

- Stop deforestation and plant billions of trees. Forests are the best and largest natural alternative to remove and store the CO_2 from the atmosphere.

- Make changes in agricultural to reduce methane and nitrous oxide emissions. Cattle raised for meat and dairy products are a major source of emissions. Most nitrous oxides are from the use of nitrogen fertilizer.

- Use carbon capture technologies to offset those uses that cannot be fully converted to electricity including some industrial processes.

- Offset the substantial subsidies for fossil fuels, mainly the ability to discharge greenhouse gases and other pollutants into the atmosphere largely for free. Consider the gradual introduction of a carbon tax with border adjustments.

- Adapt to the unavoidable temperature increases and resulting climate changes.

- Get the incentives right. Much of trillions of dollars required will be to replace aging or obsolete facilities. If the incentives are right and essential infrastructure is installed up front, most of the investment required will be provided by the private sector.

- Plan a smooth transition. If we shut down coal, oil and natural gas supplies before we have reliable renewable energy to replace it, we will have chaos with brown outs, blackouts, and big increases in energy prices.

PREPARING TO BE SUCCESSFUL

Past successful projects such as the space program initiated by President Kennedy in 1961 are examples to follow. The National Aeronautics and Space Administration (NASA) was given responsibility for this program with the authority, staffing, and funding needed. The private sector and universities were included with contracts for specific tasks. This effort was a success and completed on schedule with the moon landing in 1969.

Successful implementation of any plan to reduce emissions requires a well-thought-out overall plan covering all the major projects involved and their estimated costs and schedules. An organization must be in overall charge of the program with the staff and budget needed. This organization also needs the authority to get projects implemented and to eliminate barriers and solve problems that delay implementation or needlessly raise costs.

The president is in charge of the executive branch of government that includes all the federal government agencies. Actual funding of any program would require action by Congress and would need bipartisan political support. It should be easier to get Congress to act if there was a national plan that was supported by the president and the public. The Department of Energy, for example, could be given overall responsibility for this program. This is a large organization with much of the expertise required and with 17 national laboratories.

A STABLE ELECTRIC GRID

A specific task should be to define the stable national electric grid we need to get renewable energy from where it is best produced to where it is needed. This effort could be assigned to the Federal Energy Regulatory Commission, working with the seven independent grid system operators and regional transmission organizations across the country that are responsible for specific geographic regions.

There are many universities and national laboratories with excellent research and engineering resources that are already working toward solutions to global warming, renewable energy, and related problems.

Finally, there is a long list of private sector companies that could be tapped as prime and subcontractors. Many of these organizations are already working on renewable energy, battery technology, and other aspects of the solution.

100 PERCENT RENEWABLES SOLUTION

What electrical generation and distribution system would we need to replace fossil fuels with renewables? A common criticism is that renewable energy, solar and wind, is intermittent. There is a limit to how much renewable energy can be absorbed by the electric grid before it becomes unstable. This is an engineering problem to be solved, not a reason to limit the use of renewables.

There are several clean energy models developed by the National Renewable Energy Laboratory (NREL) and other organizations. The NREL system expects that electricity prices could be comparable or slightly lower than today's prices along with a 90 to 100 percent reduction in carbon dioxide emissions. Using the NREL model as an example, what would a system based upon 100 percent renewables look like?

A national interregional transmission system is needed with two to three times the capacity of today's transmission system to accommodate electric vehicles, heat pumps, AI data centers, greater electrification of industrial processes, and other increases in demand. This system would connect the best areas for solar and wind electricity generation with demand centers across the country. This would be a modern grid capable of forecasting and matching supply and demand. Some load shifting would be included to vary demand as well as supply in order to keep the system in balance.

Electricity source	Approximate mix (percent of total capacity)
• Solar	44%
• Wind	27%
• Nuclear	5%
• Hydro	5%
• Batteries	20% of peak capacity
• Dispatchable clean energy such as from hydrogen or pumped storage	10-15%
• Fossil fuels with carbon capture	3-5%
• Gas turbines for backup only	Less than 1%

Table 9.1: Typical grid configuration to support renewable energy

Although the specific mix of electricity sources would vary by country, Table 9.1 shows typical values for a 100 percent renewable system.

IMPORTANCE OF ECONOMICS

To make a successful transition, advocates for net zero cannot promote green energy no matter what the cost. Economics must be considered when pursuing a desirable objective. If the economics don't make sense, the transition to renewable energy and other changes needed to eliminate human-caused greenhouse gas emissions will not happen. Politicians cannot go on spending unlimited amounts of borrowed money because it is for a good cause.

The economics have to be favorable and support the investment and other costs involved. Longer term, we cannot depend upon mandates, tax breaks, and subsidies. After an initial period, subsidies and tax breaks should be phased out when it is time for a company or technology to become competitive in the marketplace. These incentives should not be allowed to become entitlements such as the ethanol subsidies that have been in place for over 40 years.

CAN WE AFFORD THE INVESTMENT REQUIRED?

One of the biggest criticisms of efforts to stop global warming is that it will cost too much. We have to sacrifice economic growth and living standards to make the transition to green energy and to eliminate all human-caused greenhouse gas emissions. Another major criticism is that the transition to renewable energy, mainly solar and wind with battery backup, will result in higher energy prices, unreliable energy supplies, and possible energy shortages. The massive effort required to make this transition has to be done without trashing the economy, increasing energy prices, or leading to unreliable energy supplies.

Estimates of the investment required to make the transition to clean energy are huge, from $75 to $150 trillion. If the transition has to be accomplished by 2050 to keep global warming under 1.5ºC, then massive annual investments have to be made—as much as $5.0 trillion/year. This is clearly unrealistic. Spending these amounts in a short time would be a crash effort leading to inefficiency and waste.

This transition will take longer. Global energy investment in 2025 is estimated to be $3.3 trillion with 66 percent, $2.2 trillion, directed toward renewable energy (solar and wind, battery storage, nuclear, the electric grid, and electric vehicles and charging stations).

If this 2025 investment of $2.2 trillion grows as fast as the global economy, at least 3.0 percent/year, then total investment in renewables over the next 30 years would be about $100 trillion ending with an annual investment in renewables of over $5.0 trillion/year (constant 2025 dollars). Based on this simple analysis and current trends, it can be assumed that funds can be available over time.

Much of this spending will be to replace aging equipment that is obsolete now or will become obsolete over the next 30 or more years. Much of the grid is old and was not designed for today's requirements.

The government will have to invest up front in essential infrastructure that facilitates private sector investments in renewable energy and in other actions to reduce emissions. The biggest incentive would be the elimination or revision of the many rules and regulations that add to the cost and time required to complete large projects, including domestic mining of essential materials from rare earths to uranium. If the incentives are right and bottlenecks are removed, most of the investment required will be provided by the private sector. If the electric grid cannot support new renewable energy projects, these projects will be delayed or cancelled. If the grid cannot support enough new charging stations, the sale of electric vehicles will be curtailed.

Does the government even need to build charging stations if grid connections are available? It didn't have to build gas stations. Tesla has already installed a national network of 2,800 supercharger stations and 30,000 fast-charging stations to support Tesla electric vehicles. If grid connections are available and as electric vehicles become more common, we can assume that shopping centers, restaurants, places of employment, apartment complexes, and others will add charging stations to meet demand. According to Inside EVs, "America is on track to add 16,700 public fast-charging ports by the end of 2025, which would be about 2.4 times the number of ports added in 2022. If this pace continues, the U.S. will have 100,000 public fast-charging ports by 2027."

WILL RENEWABLE ENERGY COST TOO MUCH?

Some critics claim that it is impossible or unaffordable to replace fossil fuels with renewables. This ignores the improving economics of renewables. The problems are real but are solvable, not reasons to abandon efforts to reduce greenhouse gas emissions. Also, the rapid progress being made in China, parts of the U.S., and in other locations demonstrates that the transition to renewables can be done and is underway.

Electricity from solar and wind farms is the cheapest form of energy in most locations. However, the cost of battery storage has to be added. Even with battery storage, new solar and wind farms are cost-competitive with electricity from new gas-fired power plants in many locations. Cost trends are in the right direction for renewables. The cost of solar panels, wind turbines, and batteries continues to decline and performance is improving. The power plants and internal combustion engines that use fossil fuels are all mature technologies with little or no opportunities for improvement.

To facilitate the switch from fossil fuels to electricity from renewables, it is essential to keep electricity rates low. Electricity rate increases are not just related to the cost of producing electricity, which should benefit from the declining costs of solar and wind farms and battery storage. The biggest factor that increases electricity prices is the need to expand and modernize the aging electric grid, transmission and distribution. Most of these grid upgrades need to be done anyway.

THE ACTION PLAN

The following is a review of the steps in the action plan:

1. **Critical need to improve electric grid reliability.**

   One of the biggest challenges in reducing dependence on fossil fuels is the need to modernize and increase the coverage, capacity, and reliability of the electrical grid—the world's electrical transmission and distribution systems. Current systems, based on 100-year-old designs, are largely "dumb systems" designed for the one-way flow of electricity from power generating stations to consumers.

   Expanding the grid needs to precede increasing demand. Electricity from renewable sources must replace gasoline and diesel fuels in transportation, natural gas for space and water heating, and coal and natural gas for electricity generation. This requires that transmission systems have good connections from sources to consumers.

Local distribution systems must have enough capacity to reliably deliver electricity as users charge more electric vehicles and transition to heat pumps for space and water heating.

Electrical loads vary during the day, day-to-day and month-to-month, depending on consumers' requirements, making the stability and reliability of transmission and distribution systems critical. Power plants must increase or decrease their output to match supply with demand. Utilities should be allowed to adjust demand by time-of-day pricing and the ability to shed non-essential loads when demand peaks to avoid brown and blackouts. Large batteries are needed to help match supply and demand, and storage systems need to be expanded.

Another challenge is public opposition to the siting of new transmission lines and solar and wind farms in many locations.[62] Large solar and wind farms should not be sited on scarce farmland or locations unacceptable to the public. They should be placed at acceptable locations with the best potential for wind and solar energy generation but must be connected to the grid with new transmission lines.

In the U.S., there are many regional and state organizations in place that should facilitate planning, expanding, and operating the transmission systems. Likewise, internationally and in Europe, codes and guidelines exist for the operation of the electricity and gas sectors.

Distribution systems are local and controlled by state regulations and organizations such as public utility commissions. The capacity of distribution systems must be substantially increased, perhaps by 300 percent or more over time, to handle the electrical load as more people charge their electric vehicles and replace natural gas with electricity for space and water heating, cooking, and heat and power for industry.

2.   Continue to increase energy efficiency and conservation.

Energy conservation and efficiency can improve from the recent 1.0 percent per year to 2.0 percent/year. The additional 1.0 percent increase will further reduce energy demand by 18 percent by 2050.

There are a number of indirect energy savings that result from the switch to renewables, including no energy to mine or process fossil fuels, cheaper energy transmission compared to pipelines and tankers, and savings due to avoiding waste heat when generating electricity. When fossil fuels are used to generate electricity, about 60 percent of the energy input is lost as waste heat eliminated in cooling water, radiators, stacks, and exhausts.

A 1.0 percent increase is a conservative and achievable estimate. Since the Arab Oil Embargo of 1973, there have been orders of magnitude of efficiency improvements for refrigerators, microwaves and other appliances, LED lights, along with the development of new building codes.

3. **Produce all electricity using renewables—primarily solar and wind.**

A growing percentage of the world's electricity is produced by renewables: mainly solar and wind, with some contribution from hydro and nuclear power. The world must replace coal with solar and wind for electricity production. This is an urgent priority. In 2021, emissions from coal-fired power plants hit a new record, exceeding 15 $GmtCO_2$. Much of this came from new coal plants in Asia that have decades of operating years ahead. While coal is declining in most of the world, it is increasing in China and India, two of the world's largest economies.

Natural gas needs to be replaced eventually. Natural gas is a cleaner fossil fuel than coal, so replacing natural gas is a lower priority.

4. **Replace fossil fuels with electricity wherever possible.**

Electricity from solar and wind could displace the use of fossil fuels as a source of heat and power for most industrial, commercial, and residential applications. As the price of fossil fuels rises (with or without the possible help of a carbon fee) and the cost of renewable energy continues to decline, more electrification will occur, increasing the demand for electricity. Residential and commercial buildings use natural gas and some oil for heating, cooking, and other needs. Most of these uses could be converted to electric power and all new construction should be designed with this in mind. Heat pumps should be used instead of furnaces, microwave ovens for cooking, and building insulation and fenestration should be improved. Some coal, oil, and natural gas are used as a raw material for the production of steel, cement, plastics, chemicals, fertilizers, and some other products. Those uses that cannot be converted to electricity have to be offset, such as by carbon capture and storage.

5. Transition transportation to electric vehicles.

About 95 percent of transportation uses liquid, oil-based fuels. This includes light and heavy vehicles, airplanes, ships, and most railroads. The biggest potential environmental improvement would be increasing the use of hybrid and electric vehicles. The use of electric vehicles is only a benefit if the electricity is produced by solar, wind, hydro, and nuclear energy. Some fossil fuel use in transportation cannot be displaced by renewable electricity, for example liquid fuels used by airplanes and oceangoing ships. However, large, all-electric ferries are already operating in Scandinavia, Canada, and the U.S.

The number of motor vehicles is forecast to increase from about 1.2 billion today to about 3.0 billion in 2050. With population growth and improved living standards, one of the first purchases is a motor vehicle, from a scooter to an automobile. Electric vehicles will increase electricity demand, and not all of this electricity will be produced by renewables. Fuel efficiency should be increased so that the remaining vehicles use less fuel per vehicle. Since vehicles last a long time, ten or more years, it will take time to replace the current vehicle fleet with new electric vehicles.

6. Reduce emissions from methane, nitrous oxides and fluorinated gases.

Methane makes up about 90 percent of the natural gas in pipelines and about 95% or more of LNG. The use of natural gas and LNG is increasing globally because it is the cleanest burning fossil fuel with the lowest carbon dioxide emissions. However, methane is also a powerful greenhouse gas and methane released to the atmosphere adds to global warming. Methane leaks account for about 30 percent of methane released into the atmosphere. This source will increase with the greater use of natural gas as a fuel. Another 30 percent of methane is accounted for by agriculture, mainly cattle-raising for meat and dairy products.

Nitrous oxides are also potent greenhouse gases. About 70 percent of nitrous oxides are from agriculture, with the use of nitrogen and organic fertilizers and the disposal of crop residue and manure. Only about 5 to 7 percent of nitrous oxide emissions are due to fossil fuel uses. Reducing nitrous oxide emissions depends upon more efficient use of nitrogen fertilizers in agriculture.

7.   Improve land use planning.

The U.S. has great potential for renewable energy from solar and wind farms—more than enough to produce all the energy for needs now and in the future, and not just electricity. Solar and wind farms should be placed in optimal locations with the most sunlight and wind potential. Locations can be developed without using scarce farmland or areas unacceptable to the public. According to a report from the National Renewable Energy Laboratory, roughly 22,000 square miles of solar panels would provide all the electricity used in the United States.[63] This is less than 1 percent of the land area of the continental U.S.

There are complaints that solar farms use more land than, for example, a gas-fired power plant. These comparisons typically compare the total area of a solar farm to the footprint of a power plant without including all the land used for oil and gas production, pipelines, storage facilities, and other land uses associated with fossil fuels. According to the National Renewable Energy Laboratory (NREL), the land needed by 2035 for 100 percent clean energy electricity production by wind and solar is a little more than the land area occupied by railroads (18,500 square miles), only about 55 percent occupied by active oil and gas leases (40,500 square miles) and only 37 percent of the land area used to grow feedstock for ethanol production (59,500 square miles).[64]

In the U.S., there is substantial land that is well suited for solar and wind. The National Renewable Energy Laboratory (NREL) published maps for the best locations for solar and wind. The best areas for solar have strong solar irradiance (sunlight), low cloud cover, affordable land, and policies that support solar energy. Some of the best areas include large sections of the southwest including desert areas in Southern California, Arizona, Nevada and New Mexico. South and west Texas are also excellent regions for solar as well as Florida.

Top locations for wind farms need consistently high and steady winds, large land areas and policies that support wind farms. This includes the central U.S., a wind corridor, which includes Kansas, Nebraska, Oklahoma, and North and South Dakota. West Texas has excellent wind potential as well. Other areas include mountain passes in California and parts of California's long Pacific coast. The East Coast from New England to the Mid-Atlantic states, the Great Lakes, and the Gulf Coast have good potential for offshore wind farms.

8.   Develop fossil fuel alternatives.

Synthetic fuels will be needed for uses that cannot be converted to electricity such as airline travel and ocean shipping. There will be a need for synthetic fuels made from recycled carbon dioxide and electricity from renewables. Synthetic liquid fuels can be produced using carbon dioxide removed from the atmosphere. It has been demonstrated that fuels that are drop-in replacements for gasoline, diesel, and aviation fuel can be produced. The challenge is to produce these fuels in large quantities at competitive prices.

Like biofuels, synthetic fuels don't actually remove carbon dioxide from the atmosphere, since the process recycles the carbon dioxide into fuel that will release CO_2 back into the atmosphere when it is consumed. The benefit is that it avoids adding additional carbon dioxide if fuels produced from oil are used. This only works if the entire process is powered by electricity from renewable sources.

Hydrogen (H_2) could become an alternative fuel as well as a means of storing energy from renewables. Hydrogen can also be transported by pipelines. It is a gas that has to be compressed to reduce its volume for transportation and storage. Hydrogen can also be used in fuel cells that are very efficient (about 60 percent) and convert hydrogen into electricity without combustion. Because there isn't any combustion, the nitrous oxide produced in internal combustion engines and power plants is eliminated. The biggest barrier to using hydrogen is its cost. For hydrogen to be a price-competitive fuel, it requires low-cost electricity from renewables. Green hydrogen is produced using electricity from renewables to produce hydrogen by electrolysis, separating water (H_2O) into hydrogen and oxygen that is discharged into the atmosphere or used in chemical processes.

Ammonia (NH_4) has potential as a liquid fuel. It can be produced using hydrogen and nitrogen from the atmosphere. Today, it is widely used primarily as a nitrogen fertilizer. The production, storage and transportation of ammonia are established technologies. Green ammonia, like hydrogen, must be produced using electricity from renewables. Ammonia does not include any carbon atoms and does not release CO_2 during combustion.

9. **Stop deforestation and plant billions of trees.**

Deforestation contributes to global greenhouse gas emissions by reducing the amount of CO_2 captured by plants through photosynthesis. The carbon is absorbed by the living plants and trees, and the oxygen is released into the atmosphere. Deforestation is largely the result of land being cleared for agriculture, particularly to increase grazing land and raise feed for cattle. The earth's plant life, mainly forests, absorbs about 30 percent of CO_2 emissions. The destruction of forests reduces the amount of CO_2 they absorb, leaving more CO_2 in the atmosphere and oceans.

There is potential for large-scale reforestation, planting billions of trees, to reduce greenhouse gases. Planting new trees requires large land areas that have soil, water, and climate to support new growth. Much of this land is in developing countries in Africa, South America, and Southeast Asia.

There is a total land area of 3.5 million square miles (roughly 110% the continental U.S.) that is suitable for planting new forests. This could increase forested areas by one-third without taking land from agriculture. This area could support an estimated 1.0 trillion new trees, compared to a total of about 3.0 trillion trees in the world today. This new growth could sequester around 24 $GmtCO_2/yr$ as the trees grow to maturity over 50 or more years. [65]

10.   Make changes in agriculture.

These emissions will be difficult to reduce. About 38 percent of the earth's total ice-free land area is currently used for agriculture. Most of the agricultural land is pastureland for livestock and the remainder is for crops. Urban areas cover only a small fraction of this land area. Agriculture is the largest source of methane and nitrous oxides. Methane is associated with raising cattle for beef and dairy products and with growing rice. Agriculture uses large amounts of gasoline and diesel as fuel and as energy for food processing. The use of nitrogen fertilizers is the main source of nitrous oxides. Reducing beef and pork consumption would reduce methane emissions. About 30 percent of methane and about 70 percent of nitrous oxide emissions can be attributed to agriculture.

Emissions from agriculture are likely to continue to increase by about 1.0 percent/year, roughly in line with population growth. It will be difficult to reduce greenhouse gas emissions from agriculture due to population growth and rising living standards that overwhelm improvements in agriculture and increase the pressure to clear more forests for agriculture. Diets change as living standards improve and people consume more meat and dairy products, requiring more land and feed for cattle. Farming methods could be improved to reduce fertilizer use and methane emissions.

11.   Carbon capture technologies.

Removing CO_2 from the atmosphere or power plant exhaust is referred to as carbon capture and storage (CCS). Removing CO_2 from the atmosphere is referred to as direct air capture (DAC).

Some amount of carbon capture and storage is needed to offset human-caused greenhouse gas emissions that cannot be eliminated. To be beneficial, the carbon dioxide removed from the atmosphere must be securely stored forever or converted into an inert material such as limestone. Some CO_2 is sequestered in trees and plants or dissolved in the oceans. If the carbon dioxide eventually leaks back into the atmosphere, then it shouldn't be removed in the first place.

In addition, the CO_2 removal process must be powered by electricity from renewables. If electricity from a fossil fuel power plant is used, the CO_2 emissions from the power plant would more than offset the CO_2 removed from the atmosphere.

12.   Offset the substantial subsidies for fossil fuels.

In Chapter 5, we describe the heavy subsidies received by fossil fuels. A fee or tax on carbon is one way to "even the playing field" for renewable energy to compete with fossil fuels. It would not be politically viable to suggest a carbon tax in the U.S. during the current administration. Globally, a number of countries have implemented some form of carbon fee. A carbon fee would also need a border adjustment tax or similar mechanism to penalize imports from countries that don't have a carbon tax. Today, it is unlikely that a global agreement could be reached on a universal carbon tax with penalties for those who don't comply.

13. Adapt to higher global temperatures.

Already, there is a large and growing need to adapt to the effects of global warming resulting from past and future greenhouse gas emissions. There will be practical consequences such as the cost and availability of property insurance in areas having a high risk of flooding, wildfires, and storm damage.

Some examples of adaptation measures are as follows:

- There will be a need to construct sea walls and take other actions to protect coastal areas. Some locations on the coast will have to be abandoned. Building in flood plains and other vulnerable areas should not be allowed.
- A greater effort will be needed to conserve water in drought-prone areas such as the Southwest in the U.S.
- Buildings will have to be upgraded to be better insulated and more heat resistant, and to replace furnaces with heat pumps. Air-conditioning will have to be added or expanded to accommodate higher temperatures.
- Rooftop solar panels with battery backup should be added where the cost can be justified.
- More facilities will be needed to protect those who cannot afford air-conditioned homes during heat waves.
- Public infrastructure must be upgraded to accommodate more severe storms.

14. Get the incentives right.

Much of the trillions of dollars required will be to replace aging or obsolete facilities. If the incentives are right and essential infrastructure is installed up front, most of the investment required will be provided by the private sector. Government loans and tax credits may be appropriate in some cases.

15.   Plan a smooth transition.

Dealing with global warming involves decreasing and eventually eliminating the use of fossil fuels. Their availability is an important part of modern economies and lifestyles. An abrupt cutoff would lead to a global recession as well as food and health crises. The companies supplying fossil fuels are large businesses globally and employ large numbers of people. This transition can take place over 30 or more years with time to adjust, for new sources of energy to be developed and put in place. There would be many job-creating opportunities associated with the transition to a more energy-efficient economy driven by electricity from renewable sources, along with investments in new science and technology. Lifestyles could change in a positive manner if low-cost solar and wind power became available in underdeveloped areas and countries that today lack access to affordable and plentiful sources of energy.

A FORECAST OF WHAT SHOULD BE POSSIBLE

The 15 actions listed above are what needs to be done to reduce greenhouse gas emissions over the next 30 or more years. It is very unlikely that, on a global basis, it will be possible to eliminate greenhouse gas emissions and stop global warming in this century.

Even if we can't see today how best to close the gap to get to net zero, we have to start reducing greenhouse gases immediately, and as fast as possible, consistent with avoiding a global recession or severe energy shortages. There will be time to explore additional actions and make more realistic estimates of when net zero, the elimination of human-caused greenhouse gases, could be achieved. In formulating a plan, it is important to set realistic goals that are actually achievable. There is no sense in launching an effort that has no chance of succeeding, even if it sounds good at the start. Unrealistic and unachievable plans that fail will destroy public confidence and support..

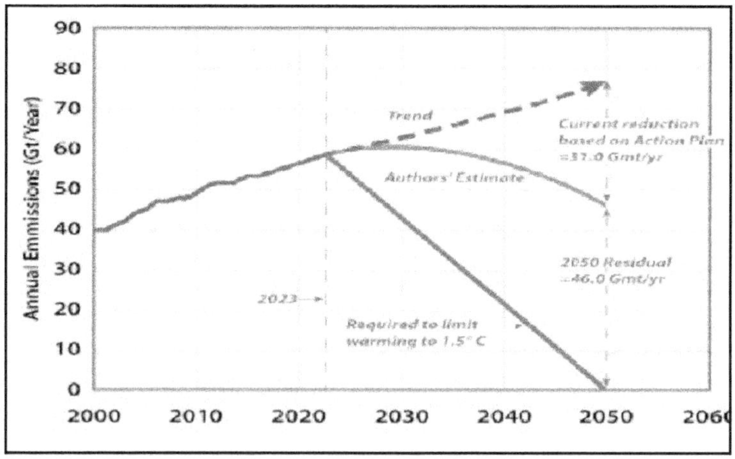

Figure 9.1: Impact of Action Plan on "Business as Usual"

Although the world will not be able to reach net zero by 2050, the IPCC goal to keep the global temperature increase under 1.5°C (2.7°F), it should be possible to avoid the trend we are on and to see greenhouse gas emissions peak and start to decline sometime in the next 20 years or so. If so, emissions will be heading in the right direction, down, even if we cannot forecast if and when net zero can be achieved in the future

The authors have made detailed forecasts of the possible impact of the measures described above. Our findings are summarized in Figure 9.1. We modeled the global trend in greenhouse gas emissions and evaluated the possible impact of accelerated conservation and efficiency measures, while estimating the effect of more rapid expansion of renewable energy forms and other measures mentioned above. Details are provided in our recent publication.[66] Our conclusion is that by 2050, it is possible to reduce global greenhouse gas emissions by about 31 billion metric tons per year, or to the level of 2010. This is still a long way from net zero, but it is moving in the right direction.

You might wonder why we don't just suck carbon dioxide out of the atmosphere to reduce the earth's temperature as some propose?

There are ways to remove carbon dioxide from the atmosphere. It sounds like a good idea, but it is not practical to remove carbon dioxide in the amount needed to make a difference.

Carbon Engineering is a leader in carbon removal technology. They estimate that it will cost $100 to $200 per ton to remove carbon dioxide from the atmosphere. The cost could be higher. Using their estimate of $200 per ton, it would cost about $8.0 trillion to remove one year's worth of human-caused emissions, about 40 gigatons of carbon dioxide.

This amounts to $40 GtCO_2 \times 1,000,000,000 \text{tons/Gt} \times 200\$/\text{ton} = \$8$ trillion to remove one year's worth of carbon dioxide emissions! This cost is over 25 percent of U.S. GDP.

In addition, it would take about 40,000 TWh of electricity to provide the power and heat this removal process would require. This is about 1.33 times all the electricity the world produces in a year.

The conclusion should be that it is much better to stop putting carbon dioxide into the atmosphere and forget about attempting to lower the earth's temperature by taking it out.

10

IMPORTANT TRENDS

Although the world is not moving fast enough, or on a large enough scale to slow down much less stop global warming, there are a number of positive trends. First, China, the number one emitter of greenhouse gases is rapidly expanding its wind and solar generating capacity and use of electric vehicles, and they project a reduction in emissions. Similarly, progress is being made in Europe, where a majority of the members favor a goal of a 90 percent cut in emissions by 2040. In the U.S., the world's second largest emitter of greenhouse gases, the government is cutting back federal efforts to reduce greenhouse gas emissions, but many states have ongoing programs. Russia, India, Australia, and Brazil are other countries unlikely to meet their self-imposed reduction targets.

There is a greater use of renewables—solar and wind with battery storage.

Globally, in 2024, about 93 percent of electricity capacity additions were for renewables, a growth rate of about 15 percent per year.[67] Of the 585 GW additions, about 77 percent was for solar and 32 percent was wind. In addition, 160 GWh of utility-scale battery storage was added in 2024, a 68 percent increase over the prior year. China accounts for 64 percent of the global renewable energy capacity additions in 2024.[68] In the U.S. in 2024, electricity capacity additions were about 62.8 GW. According to the U.S. Energy Information Agency (EIA) about 63 percent was solar, 23 percent was battery storage and 13 percent was wind, compared to 4 percent for natural gas. This allocation in favor of renewables is forecasted to continue.

Over the past decade, global energy investment has almost doubled, with clean energy investments becoming the largest share. Investments in fossil fuels are also substantial but is slowing in comparison to investments in clean energy. According to the IEA, global energy investment is forecast to increase to $3.3 trillion in 2025. An increasing share of global energy investment, about $2.2 trillion (66 percent) is being directed toward renewable energy: solar and wind electricity production, electricity storage, grid modernization, nuclear power, electric vehicles, low-emission fuels, heat pumps, and investments in energy efficiency. About $1.1 trillion (33 percent) is for coal, oil and natural gas.[69]

RENEWABLES ARE INCREASINGLY THE CHEAPEST FORM OF ENERGY.

Electricity produced by renewable energy sources (solar and wind) is cost-competitive with electricity produced using fossil fuels in most locations. The economics of renewables continue to improve with greater economies of scale and technology improvements. Fossil-fueled power plants are very mature technologies with little or no potential for improvement. Fossil-fueled power plants are affected by fuel price changes. Solar and wind are fuel-free. Adding battery storage to solar or wind electricity production adds to the cost of electricity. However, even with battery storage, solar and wind are still competitive. Combined-cycle natural gas generation is the least expensive fossil fuel source of electricity.

The levelized cost of energy (LCOE) for various sources of

electricity compares the cost of different energy technologies. It is the total life cycle cost of electricity for a given technology divided by the total life cycle electricity produced, expressed as dollars per megawatt hour ($/MWh). (See Table 10.1). Levelized cost includes the capital costs, fuel costs, operation and maintenance, financing, and assumed utilization rate.[70]

It is important to ensure that electricity from renewable sources is as reliable and available as that from fossil fuel power plants. As stated earlier, a big problem to be solved is to accommodate the variable output of solar and wind power plants to match supply and demand. Adding battery storage to solar and wind farms helps by storing electricity when excess electricity is available. Other methods may also be used, such as adding gas peaking plants to increase supply to match demand when needed. This is discussed in more detail in Chapter 9: What It Takes to Stop Global Warming.

Energy type	Source	$/MWh
Renewable	Community solar	81-217
	Utility solar	38-78
	Utility solar + storage	50-131
	Wind (onshore)	37-86
	Wind (onshore) + storage	44-123
	Wind (offshore)	70-157
	Geothermal	66-109
	Nuclear (U.S.)	141-228
Fossil fueled	Gas peaking	149-251
	Gas combined cycle	48-109
	Coal	71-173

Table 10.1 Levelized Cost of Various Energy Sources

Texas and California are the U.S. regions incorporating the most solar and wind into their electricity production. In Texas (ERCOT, the Electricity Reliability Council of Texas) about 10 percent of electricity generation is from solar, and about 24 percent is from wind. For California (CAISO, California System Independent Operator), solar accounts for about 23 percent and wind for about 10 percent of electricity output. Both Texas and California are adding additional battery capacity to help stabilize their grids. Large projects in many countries demonstrate the feasibility of renewable power. (See Table 10.2)

Large Solar Power Facilities	Gigawatts
Xinjiang Solar Farm (China)	5.0
Golmud Solar Park (China)	2.8
Bhadla Solar Park (India)	2.2
Noor Abu Dhabi (UAE)	1.2
Solar Star I & II (USA, California)	1.6
Topaz Solar Farm (USA, California)	0.6
Western Downs Green Power Hub (Australia)	0.4
Large Wind Power Facilities	**Gigawatts**
Gansu Wind Farm (China)	8.0
Alta Wind Energy Center (USA, California)	1.6
Muppandal Wind Farm (India, Tamil Nadu)	1.5
Golden Plains Wind Farm (Australia)+300 MW battery	1.4
Shepherds Flat (USA, Oregon)	0.8
Roscoe Wind Farm (USA, Texas)	0.8
Fântânele–Cogealac (Romania)	0.6

Table 10.2: Examples of Large Wind and Solar Plants

ENERGY EFFICIENCY CONTINUES TO IMPROVE

Energy conservation and efficiency can improve from the recent 1 percent per year to 2 percent per year. An additional 1 percent increase will further reduce energy demand by 18 percent over the next 25 years. We believe that 1 percent is a readily achievable estimate. Substituting electricity for fossil fuels improves efficiency substantially.

TECHNOLOGY CONTINUES TO IMPROVE

The rapidly increasing use of solar panels, wind turbines, battery storage, and electric vehicles improves economies of scale that lead to additional cost decreases and performance improvements. Increasing demand is leading to additional investments in technology and product design, and is bringing new, larger and more modern factories on line. This leads to further cost decreases and performance improvements that accelerated the acceptance and use of these renewable technologies. There will be ongoing improvements in solar cell efficiency and wind turbine design.

Advances in energy storage will help stabilize the grid, including solid-state batteries for electric vehicles, higher capacity utility-scale batteries, and green hydrogen for longer-term energy storage and for use as a fuel.

Smart electric grids and improved energy management will lead to a more stable electric grid that is better able to incorporate electricity produced by variable solar and wind. Improved battery technology and broader networks of charging stations on a smart grid will facilitate the conversion to electric vehicles. Buildings will become more energy efficient. Heat pumps will use electricity for heating as well as air-conditioning. More efficient lighting will reduce electricity demand. Cement and steel production can become more efficient to reduce CO_2 emissions.

Presently, the U.S. is enjoying a boom in oil and natural gas production thanks to fracking, a new way to extract oil and natural gas. The U.S. is now the world's largest oil producer and a net exporter of oil. How long will this last?

U.S. oil production set a new record of 13.2 million barrels per day (bpd) in 2024. The U.S. produces 20 percent of the world's oil ahead of Saudi Arabia and Russia. Oil and gas production and consumption in the U.S. have had large booms and busts with big changes in energy prices. In contrast, renewable energy from solar and wind is fuel free. Also, we won't run out of sunshine and wind someday.

Until about 1947, the U.S. was self-sufficient, producing all the oil the country needed. By 1970, oil production in the U.S. peaked at about 11.6 million bpd. In spite of this output, the U.S. was importing about 22 percent of the oil needed to meet its growing demand of 14.7 million bpd.

By 2005, only 20 years ago, U.S. production had fallen to only 7.9 million bpd while consumption increased. Oil imports peaked at 12.6 million bpd, about 60 percent of demand. That year, the U.S. spent over $250 billion on oil imports, about $410 billion in today's dollars.

This situation reversed starting about 2008 with the fracking revolution that dramatically increased domestic oil and natural gas production. A big help was that domestic consumption of oil has been essentially flat at about 20.0 million bpd since about 2003.

During these cycles, oil prices have fluctuated from a peak price of about $127 per barrel in 2008 when the economy was booming to as low as $15 per barrel in 2020 at the start of the Covid pandemic, and back to about $60 per barrel more recently.

Natural gas consumption and production show a similar boom trend in the U.S.

11

WHAT MIGHT HAPPEN?

The earth's temperature will continue to increase as long as additional human-caused greenhouse gases are discharged into the atmosphere. We can cause the world to heat up but not cool down. Global greenhouse gas emissions are continuing with no apparent end in sight. The world will be hotter by 2100, perhaps much hotter. There will be weather changes, and perhaps severe and unforeseen changes. Any plan to reach net zero would take time, perhaps 50 years or more. A crash effort to reach net zero is unlikely to be successful.

We are concerned that the reader will conclude that our discussion of what might happen is too pessimistic. None of the three scenarios we discuss includes a positive outcome under which we achieve net zero and stop additional global warming by 2100. Should we have at least one positive scenario under which emissions are reduced to zero in time to limit climate changes to some tolerable level?

What would a positive outcome require? The Action Plan discussed in Chapter 9 lists what must be done to achieve net zero. The U.S. and most other countries would have to take steps fairly soon to implement these actions by a specific date. Pledges under the Paris Agreement (promises) are not sufficient.

For the U.S., this would require a major change of direction to aggressively transition to renewable energy and implement other changes listed in the Action Plan. The U.S. would have to rejoin the Paris Agreement and show leadership in supporting international efforts to reach net zero.

The U.S. and other large countries could implement a carbon tax with a border adjustment to penalize those countries that do not have carbon taxes. A reasonable carbon tax might allow most or all subsidies for renewables to be phased out.

We would have to see fossil fuel use peak globally in the next ten years or so and start a slow and steady decline.

A potential positive is that China may successfully complete its efforts to electrify its economy and generate all the electricity needed using renewables by 2060 as discussed in Chapter 6. This would result in a significant reduction in global greenhouse gas emissions. China's green energy exports, including solar panels and electric vehicles, could help other countries cut emissions. Plus, it would confirm the fact that renewables can replace fossil fuels, creating a model for other nations to follow.

However, we must be realists and ask what is the probability that this would happen soon enough to make a significant difference. Based upon what we know today, a positive outcome is very unlikely. So, what might happen?

WHAT MIGHT HAPPEN?

Forecasting the earth's future temperature rise has to be based upon a number of assumptions and uncertainties. For example, will the earth's temperature rise as in the past be in proportion to greenhouse gases discharged into the atmosphere or will it rise faster for some reason? When will the use of fossil fuels start to decrease?

By 2100, the temperature increase could be as high as 3.0°C (5.4°F). Even if countries start to meet their NDC pledges (Nationally Determined Contributions) under the Paris Agreement, the estimated temperature increase is about 2.6 to 2.8°C (4.7 to 5.0°F).

We are already experiencing some irreversible changes such as more extensive heat waves, longer wildfire seasons, sea level rise, loss of sea ice and permafrost melt in the Arctic and melting of the Greenland and Antarctica ice caps. These changes will continue as long as the earth's temperature is elevated.

A WAKE-UP CALL?

There could be a climate change wake-up call, a 9/11 or Pearl Harbor-like event that is large enough and severe enough to cause the public and governments to realize that global warming is real and something has to be done to keep it from getting worse. This could be a severe drought or storm or perhaps exceeding one of the tipping points discussed in Chapter 4. Even then, we may do something, but would that "something" stop or slow down global warming?

Another possibility could be that the U.S. boom in oil and natural gas production due to fracking peaks and starts to decline as oil production did in 1970. If domestic production could not meet oil and natural gas consumption in the U.S., then fossil fuel imports would increase with a corresponding increase in the U.S.' trade balance. Under this possibility, the U.S. could increase the use of renewable energy to offset a shortfall in domestic oil and gas production.

THREE SCENARIOS.

In thinking about what could happen between now and 2100, it is useful to think about three broad possibilities:

- **Muddle through** with some reduction in emissions. Net zero eventually achieved many years in the future.
- **The problem persists** for much longer with greater temperature increases and more severe consequences.
- **A severe climate crisis occurs** as the earth's temperature increases, approaches, or exceeds 3.0°C. There is a possibility that one or more tipping points are exceeded.

The following is a brief discussion of each scenario.

MUDDLE THROUGH

There are some positives. Renewables are becoming even less expensive and more reliable and increasingly are displacing fossil fuels without major subsidies. Multiple large projects around the world are demonstrating that renewable energy is a practical and economical alternative to fossil fuels.

Already, about 66% of global energy investment, about $2.2 trillion in 2025, is being spent on renewable energy, displacing spending on fossil fuels. Global investment in renewables continues to increase. The transition to renewables accelerates.

The U.S. eventually rejoins the Paris Agreement and actively supports global efforts to deal with global warming. The U.S. federal and state governments start implementing the Action Plan in Chapter 9. Implementation is uneven but the decline of greenhouse gas emissions is clearly visible. There are delays with some difficult problems such as reducing emissions for some industrial processes, airlines, and ocean shipping. However, more research and development is focused on these problems.

After much discussion, the major economies start to support a carbon tax with border adjustments to penalize those countries that do not implement an effective carbon tax. This tax is implemented in small increments over time to give markets and consumers time to adjust.

China continues to lead with solar cells, wind turbines, batteries, electric vehicles, nuclear power plants, and high-speed rail with increasing R&D spending and improving economies of scale. Other countries start to compete with China, leading to even faster improvements and the more rapid adoption of clean energy. China's massive investment in a national electric grid that facilitates the integration of renewable energy demonstrates what other countries could do.

Some very large and innovative renewable energy projects such as the Australia/Singapore project demonstrate how Europe can access renewable energy from the Sahara Desert, for example. Several Middle East oil-producing countries become renewable energy power houses by developing their solar energy potential.

Global efforts improve with help being offered to less developed countries and those most vulnerable to climate change.

Public awareness of the dangers associated with global warming increases, leading to more support for actions by government, private companies, and individuals.

THE PROBLEM PERSISTS

Global warming proves to be too big, too complicated, too far off and requires an unachievable level of global coordination and cooperation to allow a well-thought-out global response. The world stumbles forward with successes and setbacks along the way.

The transition to electricity from renewables could lead to higher electricity prices and other problems. The declining cost of electricity from renewables is more than offset by required investments to expand and upgrade the electric grid, the cost of transitioning to electric vehicles and heat pumps, needed modifications to buildings and industrial facilities, and other costs.

CAN YOU BELIEVE WHAT SOME PEOPLE WANT TO DO TO AVOID REDUCING GREENHOUSE GAS EMISSIONS?

Because it is difficult to stop human-caused greenhouse gas emissions, a number of scary alternatives are being proposed to cool the earth. These are referred to as geoengineering. The basic idea is to reflect more of the sun's radiation back into space.

The underlying basis for all these ideas is the observation that past major volcanic eruptions have spewed ash and sulfur into the atmosphere. For example, nearly 20 million tons of sulfur dioxide were injected into the stratosphere during Mount Pinatubo's 1991 eruptions, and dispersal of this gas cloud around the world caused global temperatures to drop temporarily (1991 through 1993) by about 1°F (0.5°C). Some geoengineering examples are:

- Mimicking a volcano by spraying sulfate particles high into the atmosphere.
- Spraying salt water above the oceans to whiten low clouds and reflect sunlight.
- Thinning high cirrus clouds to allow more heat to escape the earth.
- Generating microbubbles on the ocean surface to reflect more sunlight.
- Covering all deserts in shiny material. (Maybe with solar panels?)
- Growing shinier crops to reflect sunlight.
- Launching millions of reflective balloons.
- Scattering trillions of silicon beads into the atmosphere.

One problem with most of these concepts is that, once released into the atmosphere, there is no way to control particle distribution, no way to predict the uneven effects of any change or to anticipate possible uncontrollable, negative side effects.

Problems and costs lead to greater opposition by the public and politicians who seek alternatives to actually doing something about global warming. These problems will be discouraging for the average citizen who may come to doubt the wisdom of transitioning to renewables. Those who oppose efforts to stop global warming will claim to be vindicated. Political conflicts and differences in national goals hamper progress. and politicians who seek alternatives to actually doing something about global warming. These problems will be discouraging for the average citizen who may come to doubt the wisdom of transitioning to renewables. Those who oppose efforts to stop global warming will claim to be vindicated. Political conflicts and differences in national goals hamper progress.

The U.S. does not rejoin the Paris Agreement and continues to ignore global warming despite increasing evidence of the problem. The U.S.continues to increase its fossil fuel use depending on its abundant domestic fossil fuel reserves..

Other more immediate problems take priority over global warming, such as inflation and the cost of living, financial crises due to excessive government debt, immigration, rapidly growing health care expenses, and other problems. New natural gas or coal-fired power plants are built to meet growth in the demand for electricity, further slowing progress in cutting greenhouse gas emissions.

China becomes the global supplier of choice for renewable energy, making this a major export industry. The U.S., Europe, and other countries become more dependent on China for renewable energy, other manufactured products and some technologies.

There is increasing pressure by U.S. consumers to buy lower-cost and clearly superior electric vehicles from China. Tariffs on electric vehicles are reduced. European and U.S. vehicle manufacturers lose sales to Chinese imports, leading to financial problems requiring government bailouts of domestic vehicles manufacturers and their suppliers.

SEVERE CLIMATE CRISIS

One or more tipping points are exceeded. It could be the wake-up call the world needs but does not lead to effective action. Global warming accelerates, and climate change becomes too big to ignore. Climate disasters—storms, coastal flooding, and wildfires—proliferate with unacceptable loss of life and more billion-dollar climate disasters each year. The public becomes very alarmed and demands action.

Politicians see a need to do something, but it is far too late to do what needs to be done. There is political gridlock in the U.S. and other countries. A crisis is declared and massive government programs are announced to stop global warming. These crash efforts spend a lot of money but do not do much good.

In desperation, governments resort to extreme measures such as geoengineering that are ineffective and have very negative consequences.

Some regions of the world become uninhabitable for at least part of the year due to high temperatures. Other regions suffer from food and water shortages. People become desperate and attempt to migrate to more hospitable locations. Other countries resort to more extreme measures to avoid being overwhelmed by desperate immigrants.

Those more fortunate individuals turn up their air conditioners, stay indoors on hot days, and complain about electricity prices.

12

WHAT CAN YOU DO TO HELP?

Decades have passed since NASA scientist James Hansen testified before the U.S. Senate in 1988 that global warming was real and was due largely to the use of fossil fuels. (See Charlie about James Hansen on page 143.) We cannot make up for lost time with a crash effort now or in the future. The facts presented in this book show that global warming is effectively irreversible—reversing it will take decades, perhaps a century or more. As discussed in the earlier chapters, we know what needs to be done and have made significant progress demonstrating that it can be done.

As described in Chapter 9, we propose an *Action Plan* starting as soon as possible to reduce global greenhouse gas emissions to about 46.0 GmtCO$_{2eq}$/yr by 2050, essentially the level of global emissions in 2010. This plan is a starting point, but alone it is not enough. But the trend in greenhouse gas emissions would decrease, with further reductions heading toward a possible net zero sometime in the future. Additional measures to reduce greenhouse gases is needed. If global warming spins out of control, more aggressive action will be required. One such eventuality could be the release of large amounts of methane from melting permafrost.

KEY CONSTRAINTS

The ten countries with the highest emissions account for over two thirds of all greenhouse gas emissions, starting with China and the U.S. If these large emitters do not take effective action, the rest of the world cannot make up the difference. There also is a "free rider" problem. If some countries do what needs to be done, but others do not, those that do take action bear all the cost and effort and the others get a free ride. This illustrates the importance of a carbon emissions fee or tax with border adjustments that penalize those countries that do not reduce their emissions.

In addition, most greenhouse gas emissions are largely beyond individual control. In the U.S., for example, with per capita greenhouse gas emissions of about 17.5 metric tons/year, estimates are that as little as 10 percent are under an individual's direct control.

As readers should know by now, the biggest step forward would be to electrify our economies and generate all the electricity use with renewables: solar, wind, nuclear, and hydro.

WHAT CAN YOU DO?

What are the best actions an individual can take to reduce greenhouse gas emissions? You are already taking an individual step forward by reading this book and learning more about a very complex problem. The top priority is to push for your state or country to commit to a specific greenhouse gas reduction target and deadline. For Americans, it would be to get the U.S. government to change direction, to rejoin the Paris Agreement, and implement effective policies and programs that reduce greenhouse gas emissions.

If you think this book is useful, get others to read it and be sure your library and local bookstores have copies. We pledge to dedicate any income from book sales we receive to purchase books for libraries, schools, and community groups.

Charlie suggests other individual actions on page 141.

Seriously consider backing projects that implement green energy programs, such as transmission lines improvements, solar and wind farms, and mining of essential minerals for batteries. We have sufficient space for renewable energy projects and can implement them without significant environmental damage. It is not necessary to put solar and wind farms on scarce farmland—abandoned oil fields are a better alternative.

WHAT COULD GOVERNMENTS DO?

This has been covered in earlier chapters. To be effective, action is needed at local, state, and federal levels. In the U.S. and many other countries, a high priority is to expand and improve the electricity transmission and distribution grids. Communities, states, and federal governments must work together to plan and implement national electric grid systems with storage capacity and the capability to transport electricity to where most customers are located.

Focus the government on economic incentives instead of mandates, regulations, and subsidies. If the incentives are right, and this is a key factor, the private sector will propose needed projects and provide the investment.

In the U.S., state agencies control electricity prices within their states. Follow Texas' example and adopt policies that keep electricity prices as low as possible. Higher electricity prices are a major deterrent to transition to renewable sources that displace fossil fuel use. Also, to meet the power demand of huge data centers for artificial intelligence (AI), abundant low-cost electricity is required. State and local agencies also have a significant role in regulations and procedures that delay or discourage new development projects. Roadblocks that delay or prevent essential developments in green energy projects should be reduced or eliminated.

Keep an eye on pending legislation and actively support measures that will help reduce greenhouse gas emissions.

WHAT CAN COMPANIES DO?

Keep innovating. Industry is both an innovator of energy-efficient devices and technology and a major energy user. Industry led the way after the 1973 oil embargo by developing and embracing new energy-efficient technologies. Anyone wanting proof of this only has to recall what it was like to buy a low-wattage fluorescent lamp in 1975. The local hardware store might have had two or three types. Today, at a large home improvement store, one encounters an aisle with hundreds of LED lamps, smart thermostats, occupancy sensors, and dozens of other energy-saving devices. There are energy-efficient refrigerators, air conditioners, and other appliances. More such products will be available in the future.

As a major energy user, industry must continue to adopt energy-saving methods and processes in manufacturing. Experience shows that conservation and efficiency improvements are always viable and wise investments. Some of this will no doubt be driven by rising fossil fuel costs. Reducing the energy component of manufacturing costs will make or keep products competitive. Remember that improving efficiency and conservation matters for service industries, which are a major part of many countries' economies.

NEW CORPORATE PHILOSOPHIES

An important shift would be to view the transition to renewables as an important new business opportunity and invest accordingly. Otherwise, the world may become increasingly dependent upon China for the green energy products, such as electric vehicles and

How can individuals help?

Educate yourself and others. Do not be misled by deniers or misinformation sources. The Appendix lists websites of reliable source you can use for fact checking.

As a voter:

- Lobby your state and congressional representatives to provide incentives for green energy projects.
- Be aware of the actions taken at the state and federal level and support or oppose projects or legislation as appropriate.
- If you are able, write op-eds and articles, give presentations, lead discussions. See our website for examples at https://theglobalclimatecrisis.com/book/
- Support legislation to make Congress more responsive to voter priorities with minimum qualifications for office and longer terms.

As an investor or stockholder:

- Support environmentally responsible companies and products.
- Advocate for stronger climate policies and use of green energy by vendors.

As a consumer:

- Choose hybrid or electric vehicles if extensive driving is necessary.
- Avoid or reduce air travel; opt for train or videoconferencing.
- Install a renewable energy source—solar panels, or connect to another renewable energy source.
- Improve home energy efficiency: upgrade insulation, use energy-efficient appliances and lighting, turn off appliances and lighting when not in use.
- Use programmable thermostats; lower heating and raise cooling temperatures.

Philosophical outlook:

If you're in your 70s, dear reader, lucky you. Why lucky? Odds are you will have given up your CO_2 when year 2050 rolls around. You won't be part of the over 9.5 billion humans populating the planet then, when odds are that the average global temperature increase could be from 2.5°C to 3.0°C (up to twice what has been in 2025). Why lucky? As the poet wrote, you'll be: "... in the dust, in the cool tombs..." Not so lucky, your grand kids, your great grand kids. It's time to think about them.

storage batteries and new technologies that will be critical in the future. Companies should make prudent investments in green energy projects and technology, and steer clear of conventional fuels subject to future scarcity and price escalation.

Through industry groups and individually, we should support well-thought-out government green energy programs that make economic sense. Likewise, review company operations and identify projects that will reduce corporate greenhouse gas emissions throughout the entire product life cycle, from raw materials production and transport, to manufacturing, to product sales, delivery, and servicing. Internally, electrify vehicle fleets, install solar panels on warehouses and other large buildings, reduce manufacturing power demand by enhancing efficiency and other measures.

Many companies are balking at the idea of governments being slow to act or ignoring global warming. They are taking actions that they believe will appeal to the majority of their customers. In a recent publication, Forbes magazine listed 100 of America's best corporate citizens.[71] Among the top 20 companies in the environmental awareness category, one finds Microsoft, IBM, Xerox, Cummins, and Bank of New York Mellon. The article covered 890 companies, and it also tells us about the worst 10 percent.

The idea of changing corporate responsibility is also reflected in corporate annual reports. Today, many major corporations provide annual "sustainability reports" to shareholders to keep them informed of the corporation's goals to improve energy efficiency and reduce greenhouse gas emissions. Some of the measures these companies reported included energy and efficiency improvements with specific goals for reducing energy use in facilities and operations, purchasing renewable energy, setting greenhouse gas emission reduction goals, investing in wind power projects, and improving vehicle fleet fuel efficiency.

Another positive move by major global corporations is joining RE100 (an acronym for Renewable Energy 100%).[72] Companies that affiliate with RE100 establish a public goal to meet 100 percent of their corporation-wide electricity demand from renewable sources within a specified timeframe, by 2050 at the latest. They also commit to disclose electrical use annually so their progress in meeting their goal can be monitored. RE100 began at Climate Week NYC 2014. Since then, the initiative has grown to include major companies in Europe and North America, India, China, Japan, and Australia. Membership encompasses diverse industries from telecommunications and information technology to cement and automobile manufacturing.

Dr. James E Hansen has a PhD from the University of Iowa and for many years was the Director of NASA's Goddard Institute for space studies.

In 1988, he testified before the U.S. Senate that global warming was potentially a serious problem and was caused by the accelerated use of fossil fuels worldwide. This was front page news for The New York Times. At the time, U.S. President George H.W. Bush made a campaign pledge to combat the greenhouse gas effect. Nothing happened. Thirty-seven years have passed, thirty-seven years lost that could have been put to good use to reduce greenhouse gas emissions.

SUCCESS IS POSSIBLE

More energy is essential for economic growth and rising living standards. We can see a future where there are abundant sources of affordable energy in the form of electricity from renewables. Hydrogen and synthetic fuels are possible forms of green energy *if* they can be produced at a reasonable cost using renewable energy. With strong public support for action by local governments and international agencies, it is possible to reduce greenhouse gas emissions, eliminate most air pollution, and still have the energy needed to improve living standards while feeding a larger global population.

13

FINAL THOUGHTS

Global warming and associated climate change is a severe and unique challenge to the world. The problem is growing and will only get worse with time. We can take action now or wait until more irreversible damage has been done and more aggressive actions are required. A delay also increases the chance of reaching a tipping point after which global warming could accelerate, or we have some climate catastrophe that forces the major world economies to act.

We authors, octogenarians, were born in 1938 and in 1939, the start of World War II in Europe. Perhaps the biggest trend that has shaped the world during our lifetimes is the substantial population increase accompanied by steadily rising living standards. These trends have driven economic growth and improved the lives of billons of people in spite of wars, recessions, pandemics and other setbacks.

In 1939, the world's population was about 2.3 billion. During our lifetimes, population growth has averaged about 1.6 percent per year and the population has grown to 8.1 billion today. During this same period, the average global living standard has grown even faster, about 2.8 percent per year and is almost eight times larger than in 1939 when measured in constant (inflation adjusted) dollars. Life expectancy, poverty, and child mortality have improved in many parts of the world. The results are uneven and there are still over a billion people living in extreme poverty or under the threat of displacement, disease, and death due to conflicts, underdevelopment, bad government, and other causes. Global warming and climate change are adding to their suffering.

We can remember the end of World War II, the atomic bombs dropped on Japan, and the atmospheric testing of hydrogen bombs by the U.S. and Russia.

In junior high school, we could watch the latest news about the Korean War on an amazing new invention—television.

During our first year as engineering students, *Sputnik* was launched by Russia and started the space race. The U.S. won, placing a man on the moon in 1969. We succeeded, but victory wasn't a foregone conclusion.

We completed our B.S. degrees using slide rules, not computers.

The Vietnam War with its domestic turmoil and political strife is a vivid memory. One of us served in the Navy during this war and was stationed in Washington DC during the Cuban Missile Crisis and the Kennedy assassination. One of us was involved in the renovation of the Pentagon, before and after 9/11.

We both spent part of our careers in the promising nuclear power industry and personally saw the rapid rise and eventual failure of commercial nuclear power. Failing big is possible even with the best intentions.

So far, we've seen the invention and widespread use of jet planes, television, a polio vaccine, nuclear weapons, the national highway system, space travel, intercontinental ballistic missiles, medical imaging, the computer, digital communications, the Hubble telescope, the internet, genomics, cell phones, and many other amazing and sometimes frightening things.

We were surprised and pleased that the Cold War ended without a military conflict. We've seen great progress as well as threats to our lives and the environment.

In spite of wars, disease, population growth, recessions, and other happenings, there has been progress. We can't rule out failure or even a catastrophe. However, we are optimists betting on eventual success with some big bumps along the way.

We are trying to present practical actions that must be taken to reduce global warming, knowing that the problem may be more difficult to solve than we are representing in this book. Even if what we recommend is not enough, it is an important step in the right direction, and the effort and expense will not be wasted.

We have a concern that the U.S. has lost its pioneering spirit and the will to imagine transformative projects: the transcontinental railroad, the Panama Canal, the Manhattan project, the Apollo program. A number of new laws and regulations will be needed. However, global warming can't be solved by long and complicated legislation with mandates, subsidies, and exceptions for special interests. Legislation and regulations should be kept simple and easily understandable by the public, the government agencies responsible for implementation, and those affected by them. Litigation may not help and could even be counterproductive, except for drawing public attention to polluters and establishing the need for tougher regulations. Let's concentrate on what's ahead, not what's happened, unless there was criminal intent.

To succeed, we need leadership: dreamers, builders, and innovators, not bureaucrats. We need to be united in our efforts to tackle this complex problem, not to be divided and distracted by those who disbelieve and discredit science. Most importantly, a coordinated global effort is essential to solving this difficult and threatening problem.

Craig Smith and Bill Fletcher

CREDITS AND ACKNOWLEDGMENTS

This book is dedicated to the memory of Frank W. Eastland, founder and CEO of the publishing company Publish Authority. Through his company, Frank supported the work of many new and experienced authors. He brought many innovations to the publishing process in support of his authors, including expanded distribution channels and access to all manner of books including print, e-books, and audiobooks. Frank offered sound publishing advice at the onset and lent his expertise in numerous ways. He reviewed early drafts of the manuscript and made suggestions ranging from the scope of the book to details for marketing it. Sadly, he passed away on the day the editor he'd selected for this project completed her work and the manuscript was ready for publication. He will be greatly missed.

We are grateful for the professionalism and expertise of the fine staff at Publish Authority. This book would not have been possible without the guidance of editor Janie Mills, Teresa Evans, and others who contributed to its development. Their careful attention to detail shaped the final format and appearance of the book and guided it through every stage of the publishing process.

It would not be possible to write this book without the dedicated efforts of thousands of scientists in many disciplines working today and in the past. Their efforts have gathered, analyzed, and reported the information needed. They have conducted experiments to verify and explain the changes we are observing. Without their efforts, the world would be experiencing changes without understanding what is happening and without insights concerning what can and should be done to prevent adverse effects. We are all in their debt.

Also, we want to acknowledge the excellent analysis and reporting done by Berkeley Earth. They are an excellent source for the latest data on global warming with excellent graphics and clear explanations.

REVIEWERS

The following reviewed early drafts of the manuscript and offered many helpful suggestions: Chuck Bair, Architect, Business development; Kevin Daly, PhD; Marty Farahat, PhD, systems analyst, professor of music and professional musician; John Kennedy, PhD professor emeritus, Chemistry, UCSB; Kelly Parmenter, PhD, engineering consultant; Tom Russell, MD; Nancy Smith, educator; Bob Taylor, Climate writer; Anne Weber, Doctorate of Medicine; Bob Weber, BS, Physics; Bob Westerly, PhD, professor Emeritus Biology; Micah Westerly, biologist, Nancy Winter, Archeologist; Arlene Westley, PhD, Psychotherapy.

We owe thanks to many people who encouraged us to write this book and who offered helpful suggestions and critiques. We are especially indebted to the scientists, engineers, economists and lawyers who provided helpful feedback on this book and on our two earlier books: Curtis Abdouch, M.S., Cecelia Arzbaecher, PhD, John J. Berger, PhD, Paul Bjorkholm, PhD, Kevin C. Daly PhD, Jerry Dauderman, MBA, Tony Hsu PhD, Thomas Merrick M.S., Joe Genshlea, Attorney, René Malés, M.S., Tom Osborne, PhD, Russell Spencer, PhD, Richard Thompson, BSEE, MBA, Peter Fletcher M.S., Marshall Lee M.S., and Thomas Russell MD.

PERMISSIONS

To those individuals and organizations who generously gave permission to reproduce illustrations or other information from their publications, we express a special thanks. We are grateful to Peter Lewellyn, Publisher-Chemistry, Earth and Environmental Science and Jessica Mack (Elsevier, Oxford, UK) who assisted us in publishing *"The Global Climate Crisis: What To Do About It (2nd edition)* 2024; Fletcher & Smith, Elsevier, 9780443273223." We recommend this book to readers who might want to delve deeper into the science of global warming and climate change. We also thank Pinya Sarasas, Programme Officer UN Environment Programme, Nairobi; Mr. George Bilicic, Managing Director, Lazard, Ms. Alyssa Fuentes, Pew Research Center; and Mr. Niall McCarthy and *Statista*, Kelly Parmenter, PhD, Parmenter Consulting, for permission to use Charlie Oscar©, and a special thanks to Shahir Masri, PhD, University of California, Irvine, who kindly reviewed Chapter 3 for technical accuracy.

We gratefully acknowledge the support and encouragement of our wives Suzy Fletcher and Nancy Smith.

APPENDIX

1. EXTRA FIGURES BY CHAPTER

a) Chapter 1

These figures are the hard physical evidence that global warming is real. Some, but not all these figures are included in chapter 1 of this book. 2024 was warmest year on Earth since direct observations began. In Berkeley Earth's analysis 2024 was 1.62 °C above our 1850-1900 average, making it the second year above 1.5 °C.

For example, Figure 1.1 in Chapter 1 shows how the earth's average temperature has risen since 1860. The global sea surface temperature has also risen 1.2 degrees C over the same time period. See Figure below.

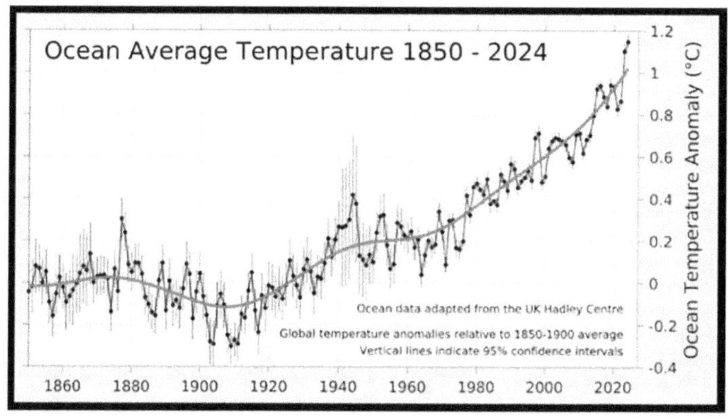

Average Global Sea Surface Temperature

The ocean temperature is rising. Oceans absorb 90 percent of the heat created by global warming. The surface temperature of the sea is defined as the first 700 meters of depth. As the ocean's temperature increases, there is more energy for storms, since hurricanes and typhoons absorb energy and moisture from the ocean.

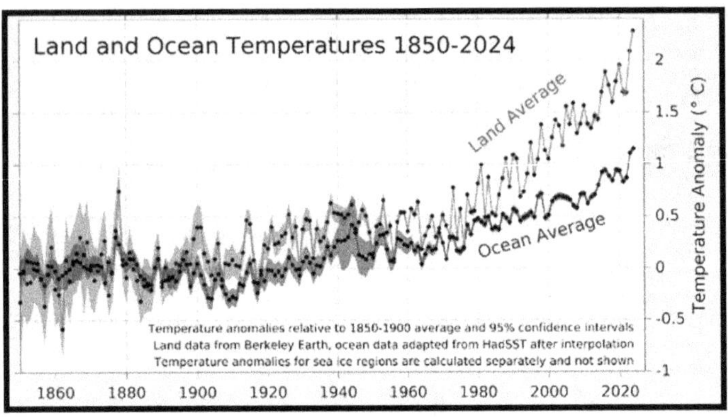

Land and sea ocean temperatures.

This figure shows that since about 1970, the land average temperature has increased faster than the ocean average temperature.

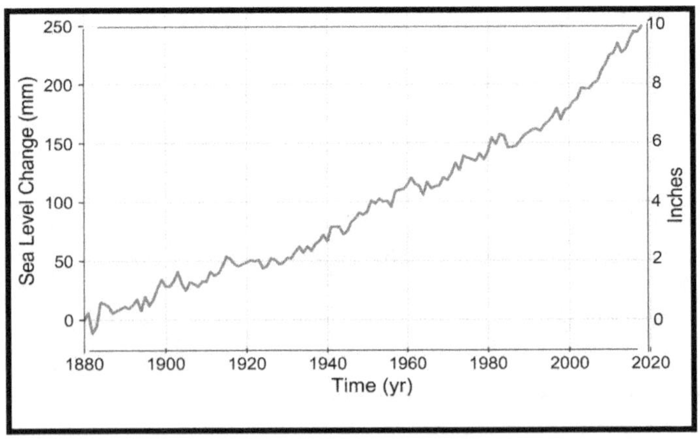

Sea Level Rise 1880-2020

Sea levels are rising. This is due to a combination of two effects: thermal expansion of warmer ocean waters, and increased meltwater from glaciers and ice caps. To date, the melting of ice caps in Greenland and Antarctica accounts for about a third of total sea level rise. The average rise so far is 21 to 24 cm (8 to 9 inches) since 1880, but can be higher or lower at specific locations. Also, the rate of sea level rise is increasing.

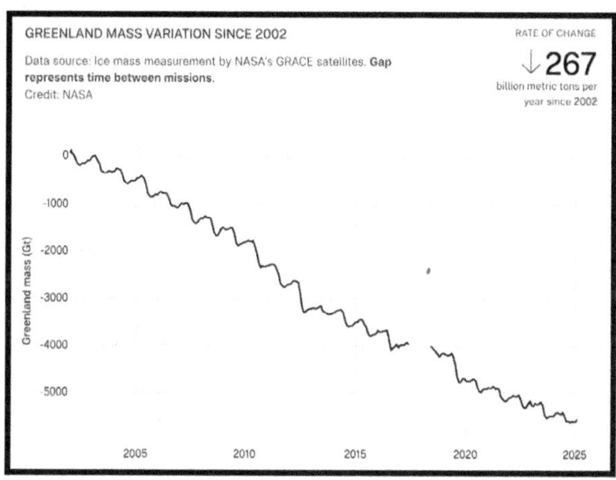

Greenland Ice Mass Variation Since 2002

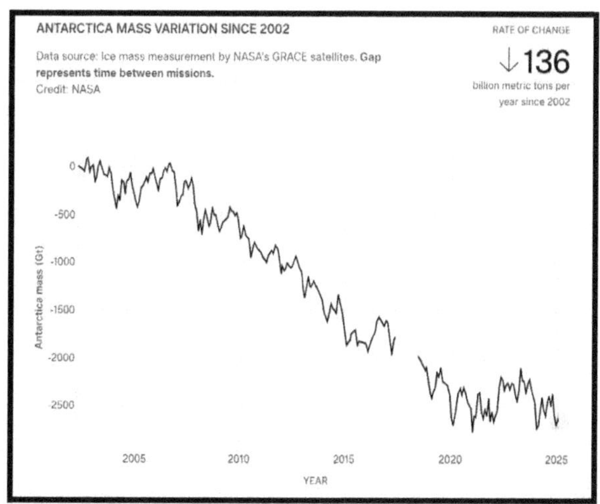

Antarctica Ice Mass Variation Since 2002

The massive ice sheets in Greenland and Antarctica are melting. In the past, these ice sheets lost ice during the summer months but replaced these losses with new snow and ice in the winter. Today, summer losses are not being fully offset. Water from melting ice caps presently accounts for about a third of sea level rise. As noted in Figures 1.5 and 1.6, Greenland has been losing 267 billion metric tons per year since 2002, and Antarctica 136 billion metric tons per year. In the Antarctic, the sheets extend out over water and break up.

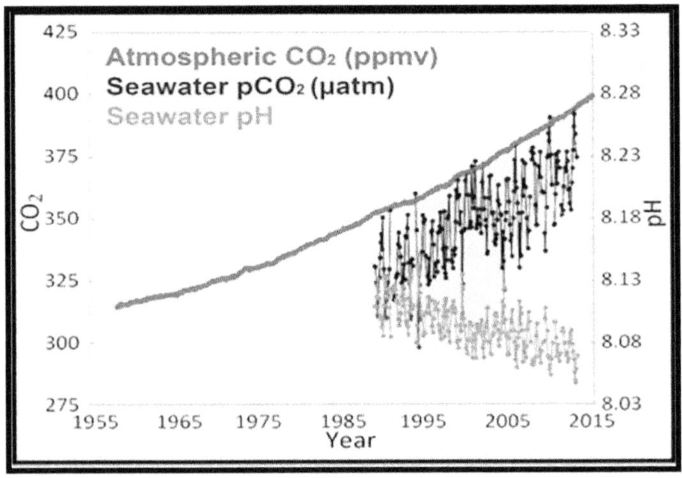

CO2 and Ocean pH at Mauna Loa, Hawaii. Ref NOAA, Our Changing Ocean

https://oceanacidification.noaa.gov/OurChangingOcean.aspx

The acidity of the oceans has also increased over the same timeframe as the atmospheric increase of carbon dioxide emissions. About 40 percent of human-caused carbon dioxide is absorbed by the oceans. It is converted to carbonic acid when it is dissolved in the ocean. Ocean acidification, a consequence of climate change, occurs when the ocean absorbs excess carbon dioxide from the atmosphere. This causes a decrease in pH and increased acidity ("pH" stands for "potential of hydrogen," a measure of acidity). pCO2 is the partial pressure of carbon dioxide in sea water, a measure of how much carbon dioxide is dissolved. This absorption helps regulate the planet's carbon dioxide levels but comes at a cost to marine life, particularly shellfish, corals, and other calcifying organisms. See Figure 1.7. The solid line shows the increasing concentration of dissolved carbon dioxide, in parts per million (volume) or ppmv. The light jagged lines show decreasing pH (greater acidity) and the dark jagged lines show increasing carbon dioxide dissolved in the ocean.

b) Chapter 3

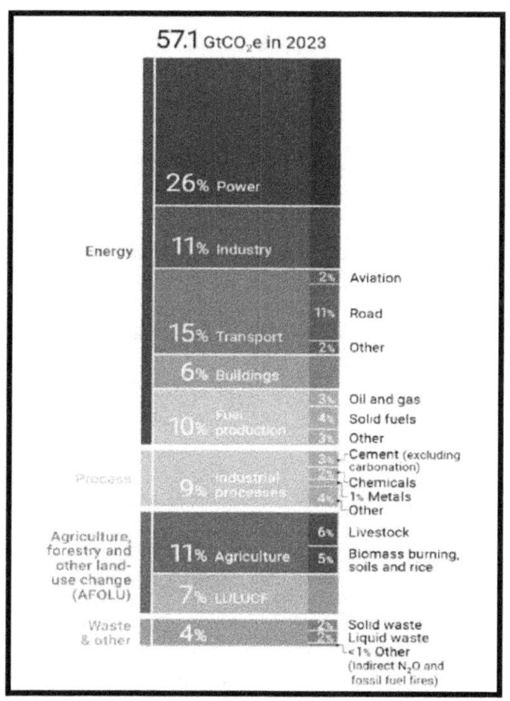

Total Global Greenhouse Gas Emissions by Sector, 2023

UN Environmental Program, October 2024. Emissions Gap Report: https://www.unep.org/resources/emissions-gap-report-2024

This figure reports total greenhouse gas emissions by source for 2024. Many other sources only report carbon dioxide emissions due to fossil fuel use which is about 75 percent of total emissions.

The largest source is power, the production of electricity that is used throughout the economy. Transportation is the second largest source and is mainly oil (gasoline and diesel fuel) used in vehicles. Agriculture, forestry and other land use change (AFOLU) measures the effects of farming and the destruction of forests, mainly to produce more land for agriculture.

c) Chapter 4

This figure produced by Climate Action Tracker is a good representation of the various estimates of forecasted temperature increases by 2100. These estimates are updated annually after the U.N.'s annual Conference of Parties (COP) meeting to update countries' pledges to reduce greenhouse gas emissions under the Paris Agreement. The 2025 COP meeting will be held in Brazil in November 2025.

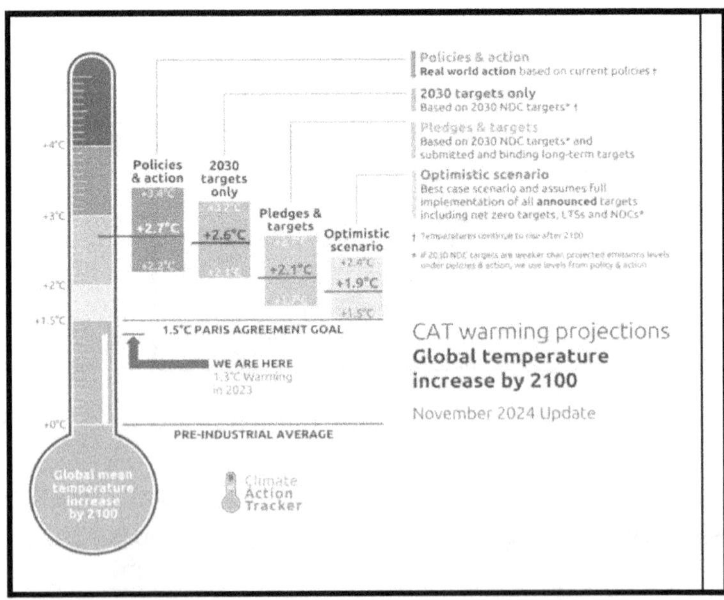

There are other authoritative estimates of global mean temperature that vary somewhat based upon various assumptions.

2. ABBREVIATIONS

GREENHOUSE GASES:

GHG (abbreviation)   Greenhouse Gas(s)

CH_4 (abbreviation)   Methane gas

CO_2 (abbreviation)   Carbon dioxide gas

CO_{2eq} (abbreviation)   Carbon dioxide gas equivalent (includes CO_2 + CH_4 and other GHG)

NOx (abbreviation)   Oxides of nitrogen

F-gases   Fluorinated gases

LNG (abbreviation)   Liquefied natural gas

COMMON MEASUREMENTS:

BTU (unit)   British thermal unit

bbl (unit)   Oil barrel

ft   (abbreviation)   Foot

GDP (abbreviation)   Gross domestic product

gal (abbreviation)   Gallon

h   (unit)   Hour

ha   hectare

hp   abbreviation)   Horsepower

I   interest rate   Dimensionless

J   (unit)   Joule

kg   (abbreviation)   Kilogram

km   (abbreviation)   Kilometer

lb   (unit)   Pound

m   abbreviation)   Meter

mpg (abbreviation)   Miles per gallon

mph   (abbreviation)   Mile per hour

toe   (abbreviation)   Metric ton of oil equivalent

y   (abbreviation)   Year

3. USEFUL REPORTS

- Fossil CO_2 and GHG emissions for all the world countries, JRC Science for Policy Report
edgar.jrc.ec.europa.eu/overview.php?v=booklet2019

- International Energy Agency (IEA): Key World Energy Statistics. www.iea.org/reports/world-energy-statistics-2019

- International Energy Agency (IEA) World Energy Outlook 2019 www.iea.org/reports/world-energy-outlook-2019

- CO_2 and Greenhouse Gas Emissions, Our World Data www.ourworldindata.org/co2-and-other-greenhouse-gas-emissions

- Global Energy Transformation: A Roadmap to 2050, International Renewable Energy Agency (IRENA) www.irena.org/publications/2019/Apr/Global-energy-transformation-A-roadmap-to-2050-2019Edition

- Global Energy System based on 100% Renewable Energy (2019), Energy Watch Group https://energywatchgroup.org/wp/wp-content/uploads/2023/12/EWG_LUT_Global100RE_power_global_GRESS_171108_final-1.pdf

- Lazard's Levelized Cost of Energy Analysis-Version 13.0 www.lazard.com/media/451086/lazards-levelized-cost-of-energy-version-130-vf.pdf

- Lazard's Levelized Cost of Storage Analysis-Version 5.0 https://www.lazard.com/media/o3ln2wve/lazards-levelized-cost-of-energy-version-130-vf.pdf

- Energy and Climate Change: Challenges and Opportunities https://www.youtube.com/watch?v=w4vtJWKF3E8

- U.S. Climate Science Special Report, Fourth National Climate Assessment (NCA4),15 Global Change Research Program (USGCRP) www.ncei.noaa.gov/news/usgcrp-climate-science-special-report

- The Climate Report. U.S. Global Change Research Program www.mhpbooks.com/books/the-climate-report/

- Climate Change Is The Greatest Threat To Human Health In History, Health Affairs Blog www.healthaffairs.org/do/10.1377/hblog20181218.278288/full/

- Climate Change in a Nutshell: The Gathering Storm www.columbia.edu/~jeh1/mailings/2018/20181206_Nutshell.pdf

- BP Energy Outlook 2019 edition https://www.bp.com/content/dam/bp/business-sites/en/global/corporate/pdfs/energy-economics/energy-outlook/bp-energy-outlook-2019.pdf

- ExxonMobil Outlook for Energy: A Perspective to 2040 https://corporate.exxonmobil.com/-/media/Global/Files/outlook-for-energy/2019-Outlook-for-Energy_v4.pdf

- The Intergovernmental Panel on Climate Change (IPCC) Special Reports: www.ipcc.ch/sr15/

- Special Report on the Ocean and Cryosphere in a Changing Climate (2019) www.ipcc.ch/srocc/

- Global Energy Perspective 2019: Reference Case. Energy Insights by McKinsey, January 2019 www.mckinsey.com/industries/oil-and-gas/our-insights/global-energy-perspective-2019

- Global Energy Perspective: Accelerated Transition. Energy Insights by McKinsey. November 2018 www.mckinsey.com/industries/oil-and-gas/how-we-help-clients/energy-insights/global-energy-perspective-accelerated-transition

- Game Changers in the Energy System: Emerging Themes Reshaping the Energy Landscape. World Economic Forum in collaboration with McKinsey & Company. January 2017 www3.weforum.org/docs/WEF_Game_Changers_in_the_Energy_System.pdf

4. USEFUL WEBSITES

- Berkeley Earth www.berkeleyearth.org/data/

- Climate Central www.climatecentral.org/

- Climate Action Tracker https://climateactiontracker.org/Carbon Clock, Potsdam Institute for Climate Impact Research

- Carbon Clock, Potsdam Institute for Climate Impact Research

- https://www.pik-potsdam.de/en/institute/departments/climate-economics-and-policy/carbon-clock/remaining-carbon-budget

- Carbon Brief www.carbonbrief.org

- Carbon Tracker Initiative www.carbontracker.org/

- Global Carbon Project www.globalcarbonproject.org/

- Our World Data https://ourworldindata.org/

- Emissions Database for Global Atmospheric Research (EDGAR) https://edgar.jrc.ec.europa.eu/

- International Renewable Energy Agency (IRENA) www.irena.org/

- Intergovernmental Panel on Climate Change (IPCC) https://www.ipcc.ch/

- U.S. National Aeronautics and Space Administration (NASA):

- Climate Change https://science.nasa.gov/climate-change/

- U.S. National Oceanic and Atmospheric Administration (NOAA) http://www.noaa.gov/

- U.S. Environmental Protection Agency (EPA) www.epa.gov/

- U.S. National Snow and Ice Data Center www.nsidc.org/home

- Copernicus Atmospheric Monitoring Service atmosphere.copernicus.eu/ https://atmosphere.copernicus.eu/

- Global Emissions Data, Center for Climate and Energy Solutions www.c2es.org/content/international-emissions/

5. TEMPERATURE CONVERSIONS

In the U.S., it is common practice to measure temperature in degrees Fahrenheit (ºF) while the more common global measurement is degrees Celsius (ºC). The following is a simple comparison between these measurements. One degree Celsius is equal to 5/9th of a degree Fahrenheit or, 1.0 ºC equals 1.8 ºF.

Fahrenheit Degrees	Celsius Degrees	Comments
212	100	Water boils
140	60	
95-104	35-40	Dangerous to work outside
98.6	37	Normal human temperature
86	30	
68	20	
59.4	15.2	Earth's average temperature 2025
56.7	13.7	Earth's preindustrial average temp
32	0	Water freezes

0.4	-18	Earth's temperature without any carbon dioxide in the atmosphere
-454	-270	Temperature in outer space

6. BIG NUMBERS

Units	Scientific Notation	Number	Prefix
Thousand	10^3	1,000	Kilo
Million	10^6	1,000,000	Mega
Billion	10^9	1,000,000,000	Giga
Trillion	10^{12}	1,000,000,000,000	Tera
Quadrillion	10^{15}	1,000,000,000,000,000	Quad

7. CONVERSION FACTORS

This book was written primarily for a U.S. audience and uses the U.S. or British measurements where possible. However, many measurements are in Standard International (SI)units commonly referred to as the metric system which is the system used internationally. The following is a summary of commonly used measures in both systems with conversion factors to convert measurements in the British system into the metric system.

Units	British or U.S.	Multiply by	To get International or Metric (SI) unit
Time	Seconds	1.0	seconds
Length	Foot	0.3048	meter
	Yard	0.9144	meter
	Mile	1.609	kilometer

Weight	Pound	0.4536	kilogram
	Short ton (2,000 lbs.)	0.9072	Metric ton (1,000 kg)
Force	Pound	4.48	newton
Volume	Gallon	3.785	liter
Pressure (lbs./in²)	Pound per sq. inch	6,895	pascal
Temperature	Degrees Fahrenheit (°F)	5/9(°F-32)	Degrees Celsius (°C)
Power	horsepower	0.746	kilowatt
Speed	miles/hour	1.609	kilometers/hour
	feet/second	0.3048	meters/second
Frequency	cycles/second	60.0	hertz
Flow (liquid)	gallons/minute	3.785	liters/minute
Flow (gas)	Cubic feet/minute	471.95	Cubic cm/second
Area	Square feet	0.0929	Square meter
	Acre	0.4047	hectare
	Square mile	2.589	Square kilometer

Power	horsepower	746	Watt
	kilowatt	0.001	Watt (one joule/second)
Energy	BTU	1,055	joule
	Kilowatt hour	3.6×10^6	joule

8. POWER VERSUS ENERGY

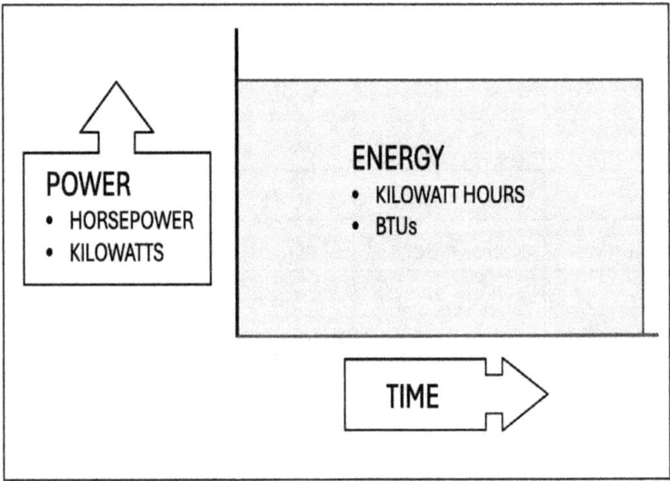

- Power is the rate at which a machine can produce energy. An example is a nuclear power plant that can produce 1,200 megawatts (MW) or a car engine rated at 250 horsepower (hp).

- Energy is the total amount of power produced over time such as megawatt hours.

There are different units used to measure power and energy, and this can be very confusing. Some of the most common units will be summarized and compared.

In discussing energy such as how much energy does the U.S. consume in a year, it is common practice to convert various forms of energy into a common measurement such as megawatt hours even for energy that is not produced as electricity. For example, a BTU (British Thermal Unit) is a common measurement for heat energy such as from burning natural gas or coal to produce heat. One kilowatt hour is equal to 3,412 BTUs.

9. ENERGY AND CARBON DIOXIDE EMISSIONS

Fuel	Energy content (common measure)	Energy Content (MJ/kg)	CO_2 emissions (kg of CO_2 emitted during combustion of one unit)
Hydrogen	61,000 BTU/lb.	120	0
Coal (anthracite)	24-26 MBTU/ton	28-30	2592/ton
Coal (bituminous)	20-30 MBTU/ton	25-35	2238/to
Crude oil	5.8 MBTU/bbl	38.5	430/bbl
Natural gas	1000 BTU/ft3	47-56	0.0548/ft3
Methane	500-700 BTU/ft3	28-38	0.05/ ft3
Diesel fuel	130,300 BTU/gal	43	10.19/gal
Gasoline	127,600 BTU/gal	48	8.78/gal
Kerosene (jet fuel)	135,000 BTU/gal	46	9.75/gal
Wood	8,000-10,000 BTU/lb.	19-23	0.834/lb.

10. AUTHORS' PUBLICATIONS

- Fletcher, William D. and Smith, Craig B., "The Global Climate Crisis: What to do about it." Elsevier, Oxford UK, 2024.

  The purpose of the 2nd edition is to provide a complete understanding of global warming with clear explanations of the science behind climate change. The book contains the facts people need to understand global warming and what can and should be done about it, along with the latest data concerning global temperatures, sea level rise, and other concerns. We discuss actions that need to be taken, including a review of relevant past successes and failures with lessons learned. The complexities and challenges of addressing global warming are discussed.

- Fletcher, William D. and Smith, Craig B., "Reaching Net Zero: What It Takes to Solve the Global Climate Crisis." Elsevier, Oxford UK, 2020. This was the authors' first book on global warming written to fully understand the causes of global warming, its effects, and possible actions needed to stop further increases in the earth's temperature.

- Smith, Craig B. and Parmenter, Kelly E., "Energy Management Principles: Applications, benefits, savings," (2nd ed.), Elsevier, Oxford UK, 2016.

  This book is a comprehensive guide to the fundamental principles and systematic processes of maintaining and improving energy efficiency and reducing waste. It provides extensive coverage of all major fundamental energy management principles. It applies general principles to all major components of energy use such as HVAC, electrical end use and lighting, industrial processes, and transportation and describes how to initiate an energy management program for a building, a process, a farm or an industrial facility.

The articles listed below are available on the authors' website
The Global Climate Crisis, www.theglobalclimatecrisis.com/

ARTICLES/WEBSITE

- "Climate Activists at 85" by Liz Galst, *Solutions, How We Protect the Planet Now,* Winter 2025
  https://library.edf.org/CS.aspx?VP3=CMS3&VF=Home

- UCSB DAILY NEXUS, Monday, May 6, 2024. "Are we fossil fools? A wake-up call for Earth Day "by Sophia Delap
  https://dailynexus.com/2024-05-02/are-we-fossil-fools-a-wake-up-call-for-earth-day/

- *Global Warming Is Irreversible.* By William D. Fletcher and Craig B. Smith, Real Clear Energy April 15, 2025
  https://www.realclearenergy.org/articles/2025/04/15/global_warming_is_irreversible_1104194.html

- *The Transition to Renewables Is Underway,* by William D. Fletcher and Craig B. Smith, Real Clear Energy, July 17, 2024
  www.realclearenergy.org/articles/2024/07/17/the_transition_to_renewables_is_underway_1045048.html

- *Is Carbon Capture Big Oil's Next Pay Day?* by Craig B. Smith and William D. Fletcher, Real Clear Energy July 9, 2024
  https://www.realclearenergy.org/articles/2024/07/09/is_carbon_capture_big_oils_next_pay_day_1043389.html

- *What It Takes to Stop Climate Change (Part 1)* by William D. Fletcher and Craig B. Smith, The Messenger October 17, 2023 The Messenger website is no longer functioning

- With Climate Change, Failure Is an Option — One We'll Have to Live With (Part 2) by William D. Fletcher and Craig B. Smith, The Messenger October 23, 2023 The Messenger website is no longer functioning

- *Can we overcome the hurdles for nuclear power revival?* by William D. Fletcher and Craig B. Smith, The Hill April 14, 2023 - https://thehill.com/%20opinion/energy-environment/3950501-can-we-overcome-the-hurdles-for-nuclear-power-revival/

- *The renewables sweepstakes* by William Fletcher and Craig B. Smith, The Hill March 17, 2023 https://thehill.com/opinion/energy-environment/3905272-the-renewables-sweepstakes/

- *Can we stop global warming?* by William D. Fletcher and Craig B. Smith, The Hill, January 9, 2023 https://thehill.com/opinion/energy-environment/3805992-can-we-stop-global-warming/

- *A Climate Change Code Red by* Craig Smith & William Fletcher, Peace Magazine October 2021 http://peacemagazine.org/archive/v37n4p12.htm

- *Are we Facing a Climate Apocalypse?* by Craig B. Smith and William D. Fletcher, Peace Magazine January 2021 http://peacemagazine.org/archive/v37n3p25.htm

11. SUPPLEMENTARY REFERENCES

This book has over 70 endnotes that will be of use to the reader. In addition, we are providing additional references here for the reader who wishes to delve deeper into the science of global warming and climate change.

- CHAPTER 1: Global warming is real.Proof that global warming is real. Oxford Academic. BioScience. The 2024 state of the climate report: Perilous times in planet Earth October 2024 https://academic.oup.com/bioscience/article/74/12/812/7808595

- CHAPTER 2: Global warming is irreversible. Human-caused emissions. U.S. National Aeronautical and Space Administration (NASA) Graphic: Greenhouse Gas Sources, Lifespans, and Possible Added Heat. (GREAT GRAPHIC) https://science.nasa.gov/resource/graphic-major-greenhouse-gas-sources-lifespans-and-possible-added-heat/

- Gases in the atmosphere to be reduced by half. Earth.org. The Time Lag of Climate Change by Owen Mulhern dated April 27, 2020 https://earth.org/data_visualization/the-time-lag-of-climate-change/

- CHAPTER 4: Dangers of global warming.Earth's climate and population.IPCC AR6 Synthesis Report: Climate Change 2023. https://www.ipcc.ch/report/ar6/syr/downloads/report/IPCC_AR6_SYR_LongerReport.pdf

- Additional global warming information. American Meteorological Society (AMS) International "State of the Climate" report confirms record-high greenhouse gases, global temperatures, global sea level, and ocean heat in 2024 dated August 14, 2025 https://www.ametsoc.org/ams/about-ams/news/news-releases/international-state-of-the-climate-report-confirms-record-high-greenhouse-gases-global-temperatures-global-sea-level-and-ocean-heat-in-2024/ .

- Effects of global warming. Effects of Human-Caused Greenhouse Gas Emissions on U.S. Climate, Health, and Welfare (2025) https://www.nationalacademies.org/read/29239/chapter/1

- A sceptic's guide to climate change.Berkeley Earth. (cause effects matrix) Know the facts: A skeptic's guide to climate change. https://berkeley-earth-wp-offload.storage.googleapis.com/wp-content/uploads/2022/12/03232411/skeptics-guide-to-climate-change-1.pdf

- Effect of temperatures on climate change. WORLD RESOURCES INSTITUTE. 10 Big Findings from the 2023 IPCC Report on

Climate Change. March 20, 2023 By Sophie Boehm and Clea Schumer https://www.wri.org/insights/2023-ipcc-ar6-synthesis-report-climate-change-findings

- Temperatures in air and oceans. Yale Climate Connections. The planet is 'on the brink of an irreversible climate disaster,' scientists warn by Dana Nuccitelli, October 2024. https://yaleclimateconnections.org/2024/10/the-planet-is-on-the-brink-of-an-irreversible-climate-disaster-scientists-warn/

- State of climate report. The American Institute of Biological Sciences. The 2024 state of the climate report: Perilous times on planet Earth by William Ripley et al dated October 2024 https://reflexus.org/wp-content/uploads/The-2024-state-of-the-climate-report-Perilous-times-on-planet-Earth.pdf

- Tipping points: potentially large, irreversible and self-perpetuating. CarbonBrief Climate tipping points have put Earth in 'disastrous trajectory' December 2023 https://www.carbonbrief.org/qa-climate-tipping-points-have-put-earth-on-disastrous-trajectory-says-new-report/

- CHAPTER 6: Renewable energy alternatives. Rated power output continuously. U.S. Energy Information Administration (EIA) Electric Power Monthly Tables 6.07.A&B https://www.eia.gov/electricity/monthly/epm_table_grapher.php?t=epmt_6_07_b

- Table 6.1 Various Electricity Generating Plants.U.S. Department of Energy. What is Generation Capacity? March 30, 2025 https://www.energy.gov/ne/articles/what-generation-capacity (FOR FIG 6.1)

- Substituting solar energy for fossil fuels. Reuters. The top global solar power potential hotspots by Gavin Maguire, May 230 2023   https://www.reuters.com/business/energy/top-global-solar-power-potential-hotspots-maguire-2023-05-23/

- Global Solar Atlas (GSA) https://globalsolaratlas.info/map?c=11.609193,8.349609,3

- Fastest growing markets for solar power. Oil Giant Saudi

Arabia Is Emerging as a Solar Power https://www.wsj.com/business/energy-oil/oil-giant-saudi-arabia-is-emerging-as-a-solar-power-e0652a56?gaa_at=eafs&gaa_n=ASWzDAh3bRcyu1r57QVVjamxcUVMxhsfdOiYx-czJ3uFdT4z6tY2QSarYipdryLbszY%3D&gaa_ts=68e547cb&gaa_sig=5cMkCwA_La9EEC0k6ORrCKMxOZrWdcjApuC-Z8S2wRNN3A9b4SgFN2J9bH32aANLT1snj4f5x29luBsf2LeK_Q%3D%3D

- Natural gas production in the US. Texas Economic Development Corporation. Texas is the Top State for Oil and Gas Production and for Wind-powered generation. February 6, 2025 https://businessintexas.com/rankings/texas-is-the-top-state-for-oil-and-gas-production-and-for-wind-powered-generation/

- Electricity production using renewables. The Motley Fool, Which States Produce the Most Renewable Energy by Lyle Daly September 2, 2025 https://www.fool.com/research/renewable-energy-by-state/

- Competitive electricity market. Texas Electricity Market Explained by Mary Pressler, July 22, 2023. https://quickelectricity.com/texas-electricity-explained/

- Wind capacity and battery storage. PVTECH. Texas outpaces California as US state with the most utility-scale PV capacity by Jonathan Tourino Jacobo. September 6, 2025 https://www.pv-tech.org/texas-outpaces-california-as-us-state-with-most-utility-scale-pv-capacity/

- Battery storage to stabilize the grid. Rabobank. Texas: A high stakes frontier for US battery energy storage systems by Amit Mathrani. July 1, 2025 https://www.rabobank.com/knowledge/d011484585-texas-a-high-stakes-frontier-for-us-battery-energy-storage-systems

- Electric grid managed by ERCOT (The Electric Reliability Council of Texas). Austin Energy: Statewide Electric Grid, A Modern Grid to Meet Modern Demand. May 5, 2024 https://austinenergy.com/about/company-profile/electric-system/statewide-electric-grid

- Competitive electricity market. Texas Electricity Market Explained by Mary Pressler, July 22, 2023. https://quickelectricity.com/texas-electricity-explained/

- Competitive electricity market. EUMMOT: Electric Utility Marketing Managers of Texas. Deregulation and Competitive Electricity Market. https://texasefficiency.com/industry-overview/

- California electricity costs 26 cents/kWh. Electric Choice: See electricity rates in your zip code dated September 22, 2025 https://www.electricchoice.com/electricity-prices-by-state/

- Electricity customers face crisis. Foreign Affairs: The Coming Electricity Crisis by Brian Deese and Lisa Hansmann dated September 9, 2025 https://www.foreignaffairs.com/united-states/coming-electricity-crisis-data-brian-deese

- Green hydrogen, a fuel for the future. Power Engineering: Texas project to create hydrogen 'ecosystem' as precursor to hydrogen hubs by Sean Wolfe. April 24, 2024 https://www.power-eng.com/hydrogen/texas-project-to-create-hydrogen-ecosystem-as-precursor-to-hydrogen-hubs/

- California close second to Texas in renewable energy. EIA: U.S. Energy Information Administration. California State Profile June 20,2025 https://www.eia.gov/state/analysis.php?sid=CA

- High rates requested by the California's three large utilities. CA.gov: Public Utilities Commission. Electric Rates https://www.cpuc.ca.gov/industries-and-topics/electrical-energy/electric-rates

- High speed rail, and nuclear power. CarbonBrief. China. Various authors. November 29, 2023 https://interactive.carbonbrief.org/the-carbon-brief-profile-china/index.html

- In 2025, over $600 billion energy investment. International Energy Agency (IEA). World Energy Investment 2025. https://www.iea.org/reports/world-energy-investment-2025/china

- New global renewable energy capacity additions. Enerdata

April 4, 2025, Global renewable capacities rose by 585 GW in 2024, with China accounting for 64%. https://www.enerdata.net/publications/daily-energy-news/global-renewable-capacities-rose-585-gw-2024-china-accounting-64.html

- World's largest oil importer in 2013. U.S. EIA. China surpassed the United States as the world's largest crude oil importer in 2017. https://www.eia.gov/todayinenergy/detail.php?id=37821

- Foreign exchange needed to pay for oil. Wall Street Journal, July 21, 2025, How China Curbed Its Oil Addiction-and Blunted a U.S. Pressure Point

- Energy will come from non-fossil sources. CarbonBrief, China will need 10,000GW of wind and solar by 2060, multiple authors, March 17, 2025. https://www.carbonbrief.org/guest-post-china-will-need-10000gw-of-wind-and-solar-by-2060/

- Big increase in China's electricity demand. CarbonBrief. Record solar growth keeps China's CO2 falling in first half of 2025. Jauri Myllyvirta August 21, 2025 https://www.carbonbrief.org/analysis-record-solar-growth-keeps-chinas-co2-falling-in-first-half-of-2025/

- Energy from non-fossil sources. National Development and Reform Commission (NRDC)Working Guidance for Carbon Dioxide Peaking and Carbon Neutrality in Full and Faithful Implementation of the New Development Philosophy October 25, 2021, https://english.www.gov.cn/policies/latestreleases/202110/25/content_WS61760047c6d0df57f98e3c21.html

- Wind (47%), nuclear power (19%) and hydro and biomass (20%).Elements, Visualizing China's Evolving Energy Mix. Bruno Venditti. July 13, 2021. https://elements.visualcapitalist.com/visualizing-chinas-energy-transition/

- China: a massive upgrade of its grid. The Progress Playbook. China is leading the way in electrification, too. Nick Hedley. June 19, 2025 https://theprogressplaybook.com/2024/06/19/china-is-leading-the-way-in-electrification-too/

- China recycles nuclear waste into new reactor fuel. World Nuclear Association. China's Nuclear Fuel Cycle. April 25, 2024. https://world-nuclear.org/information-library/country-profiles/countries-a-f/china-nuclear-fuel-cycle

- China has a high percent of the world's electric vehicles. IEA Global EV Outlook 2025 https://www.iea.org/reports/global-ev-outlook-2025/trends-in-the-electric-car-industry

- Charging station for every 2.7 electric vehicles. Argus, China expands EV charging infrastructure in 2024 https://www.argusmedia.com/en/news-and-insights/latest-market-news/2650730-china-expands-ev-charging-infrastructure-in-2024

- CHAPTER 9: What it takes to stop global warming. US NREL on renewable energy U.S. Department of Energy, National Renewable Energy Laboratory 100% Clean Electricity by 2035 Study https://www.nrel.gov/analysis/100-percent-clean-electricity-by-2035-study (for figure on page 85)

- Carbon Credits. What's Behind the $53 Trillion Energy Investment Needed for Net Zero? By Jennifer L. May 21, 2025. https://carboncredits.com/whats-behind-the-53-trillion-energy-investment-needed-for-net-zero/

- Inside EVs. America's EV Charging Network Is About To Skyrocket by Suvrat Kothari, July 28, 2025. https://insideevs.com/news/767059/record-ev-charger-deployment-q2-2025/p

- Global energy investment:$3.3 trillion with 66 percent, $2.2 trillion, International Energy Agency (IEA), Global energy investment set to rise to $3.3 trillion in 2025 amid economic uncertainty and energy security concerns, June 5, 2025 https://www.iea.org/news/global-energy-investment-set-to-rise-to-3-3-trillion-in-2025-amid-economic-uncertainty-and-energy-security-concerns

- Wind and solar cheapest form of energy in most locations. Scientific American. Wind and Solar Energy Are Cheaper Than Electricity from Fossil-Fuel Plants by Benjamin Storrow, June

17, 2025. https://www.scientificamerican.com/article/wind-and-solar-energy-are-cheaper-than-electricity-from-fossil-fuel-plants/

- Efficiency: various types of power plants in many locations. Lazard's Levelized Cost of Energy (LCOE) 2025 https://www.lazard.com/research-insights/levelized-cost-of-energyplus-lcoeplus/

- Figure 9.1 in Chapter 9: Plan on "Business as Usual" The Global Climate Crisis: What To Do About It. February 2024 by William Fletcher and Craig Smith. https://shop.elsevier.com/books/the-global-climate-crisis/fletcher/978-0-443-27322-3

- Chapter 11: What might happen WORLD RESOURCES INSTITUTE. 10 Big Findings from the 2023 IPCC Report on Climate Change. March 20, 2023 By Sophie Boehm and Clea Schumer https://www.wri.org/insights/2023-ipcc-ar6-synthesis-report-climate-change-findings

- Emission pathways. Climate Action Tracker. Emissions Pathways to 2100. November 2024. https://climateactiontracker.org/global/emissions-pathways/

INDEX

NOTES

CHAPTER 1

[1]   Berkeley Earth: https://berkeleyearth.org/global-temperature-report-for-2024/

[2]   NOAA Global Monitoring laboratory: trends in CO2, CH4, N2O, SF6: https://gml.noaa.gov/ccgg/trends/

[3]   Raw materials, plastics and fertilizers impact CO2 emissions: https://www.csiro.au/en/research/environmental-impacts/climate-change/climate-change-qa/sources-of-co2

[4]   NASA: Sea levels are rising: https://climate.nasa.gov/vital-signs/sea-level/?intent=121

[5]   Climate.gov: Sea level rise rate increasing: https://www.climate.gov/news-features/understanding-climate/climate-change-global-sea-level

[6]   NOAA says Arctic warming faster: https://arctic.noaa.gov/report-card/report-card-2024/executive-summary-2024/

[7]   Arctic warming faster than rest of planet:https://www.carbonbrief.org/guest-post-why-does-the-arctic-warm-faster-than-the-rest-of-the-planet/

[8]   National Snow and Ice Center: Is Arctic sea ice loss changing the weather?: https://nsidc.org/learn/ask-scientist/declining-sea-ice-changing-atmosphere

9   NOAA in the Arctic Report Card update for 2024: https://www.climate.gov/news-features/featured-images/2024-arctic-report-card-arctic-tundra-now-net-source-carbon-dioxide

10   JPL: NASA study: More Greenland ice lost than previously estimated: https://www.jpl.nasa.gov/news/nasa-study-more-greenland-ice-lost-than-previously-estimated/

11   JPL: NASA study: More Greenland ice lost than previously estimated: https://www.jpl.nasa.gov/news/nasa-study-more-greenland-ice-lost-than-previously-estimated/

CHAPTER 2

12   MIT Climate Portal Writing Team with Prof. Ed Boyle, "How do we know how long carbon dioxide remains in the atmosphere." MIT Center for Global Change science, January 17, 2024: https://cgcs.mit.edu/how-do-we-know-how-long-carbon-dioxide-remains-atmosphere.

CHAPTER 3

13   NASA, Global Climate Change, Graphic: The Relentless Rise of Carbon Dioxide, site last updated January 9, 2020, https://climate.nasa.gov/climate_resources/24/graphic-the-relentless-rise-of-carbon-dioxide/. Shows CO2 variation in ancient times.

14   Qiancheng Ma, "Greenhouse gases: refining the role of carbon dioxide," NASA, Goddard Institute for Space Studie s, Science Briefs, March 1998, https://www.giss.nasa.gov/research/briefs/ma_01/. Shows that without CO2, Earth's temperature would be -18° C

15   Total global greenhouse gas emissions by sector, 2023: UN Environmental Program, October 2024. Emission Gap Report: https://www.unep.org/resources/emissions-gap-report 2024./ This report has a detailed breakdown by sector (see figure in appendix) and states that total global emissions in 2023 were 57 billion GMT of CO2 equivalent.-

16   This report has a detailed breakdown by sector (see figure in appendix) and states that total global emissions in 2023 were 57 billion GMT of CO2 equivalent.-

17   World passes 30% renewable electricity, EMBER, May 7, 2024: https://ember-energy.org/latest-updates-world-passes-30-renewables-electricity-milestone

18   See Fletcher, William D. and Smith, Craig B., The Global Climate Crisis: What to do about it," pp. 68-70, Elsev ier, Oxford, U.K., 2024

19   Hans E. Suess, "Letters," (On the measurement of C-14 in the atmosphere) Science, 122, no. 316, (Sept 1955): 41 5-416.

20   Heather D. Graven, (Aug 2015). "Impact of fossil fuel emissions on atmospheric radiocarbon and various applictions of radiocarbon over this century," PNAS, 112, no. 31, (Aug 2015): 9542-9545n

21   Data is extracted from GHG emissions of all world countries-2024: EDGAR - Emissions Database for Global Atmospheric Research, https://edgar.jrc.ec.europa.eu/report_2024

22   Shahir Masri, Beyond Debate: Answers to 50 Misconceptions on Climate Change (Newport Beach, CA: Dockside Sailing Press, 2018), 31-33

23   History of the IPCC: Founding: https://www.ipcc.ch/about/history/

24   IPCC created limits of 1.5 degrees C and 2.0 degrees C to limit damage from global warming: https://www.ipcc.ch/sr15/chapter/chapter-1/

25   Global Carbon Budget: Fossil fuel CO2 emissions increase again in 2024: https://globalcarbonbudget.org/fossil-fuel-co2-emissions-increase-again-in-2024/

26   MIT Climate Portal: Why did the IPCC choose 2 degrees C as the goal for limiting global warming? https://climate.mit.edu/ask-mit/why-did-ipcc-choose-2deg-c-goal-limiting-global-warming.

27   Statista, Energy & Environment- Climate and Weather. By 2100, average global temperature could be 2.5 to 3.0 degrees C, with 2.7 degrees an optimistic scenario depending on policies and action: https://www.statista.com/statistics/1278800/global-temperature-increase-by-scenario/?srsltid=AfmBOorTKfYW98TZ6d8WCJvkvO6-8vSof5nekQlL4h7d6mCbiH1U5WhK

CHAPTER 4

28   Los Angeles Times, "This Western mega drought is the worst in a millennium," p. A1, 2-15-22

29   Climate.gov Climate Change: Global Sea Level by Rebecca Lindsey August 22, 2023 https://www.climate.gov/news-features/understanding-climate/climate-change-global-sea-level

30   Henry Fountain, "Scientists Link Hurricane Harvey's Record Rainfall to Climate Change," The New York Times, December 13, 2017

31   NASA Wildfires and Climate Change. https://science.nasa.gov/earth/explore/wildfires-and-climate-change/

32   Arctic Council. Shifting Winds: How a Wavier Polar Jet Stream Causes Extreme Weather Events. Jennifer Francis, October 28, 2024. https://arctic-council.org/news/shifting-winds-how-a-wavier-polar-jet-stream-causes-extreme-weather-events/

33   EPA: Climate change indicators: Length of growing season: https://www.epa.gov/climate-indicators/climate-change-indicators-length-growing-season#

34   NASA, Climate Change and Its Environmental Impacts on Crop Growth. https://www.nasa.gov/earth-and-climate/nasa-at-your-table-climate-change-and-its-environmental-impacts-on-crop-growth/

35   99% of coral reefs could disappear if we don't slash emissions this decade, alarming new study shows. World Economic Forum, February 4, 2022: https://www.weforum.org/stories/2022/02/coral-reefs-extinct-global-warming-new-study/

36   U.S. Environmental Protection Agency, Climate Change Indicators: Ocean Acidity, site last updated August 2016, https://www.epa.gov/climate-indicators/climate-change-indicators-ocean-acidity.

37   Karen Limburg, "The Ocean is Losing its Breath—and Climate Change is Making it Worse," SciTechConnect, November 10, 2016, http://scitechconnect.elsevier.com/ocean-losing-breath-climate-change-worse/

[38]   Associated Press, "Nature in Deep Trouble, Report Says," Los Angeles Times, May 7, 2019, p A4. See also United Nations, "UN Report: Nature's Dangerous Decline 'Unprecedented'; species extinction rates 'accelerating,'" May 6, 2019, https://www.un.org/sustainabledevelopment/blog/2019/05/nature-decline-unprecedented-report/

[39]   David Introcaso, "Climate Change Is the Greatest Threat to Human Health in History," The Lancet Countdown on health and climate change from 25 years of interaction to a global transformation for public health, December 19, 2018.

CHAPTER 5

[40]   Hannah Ritchie, Max Roser, and Pablo Rosado (2020) - "CO_2 and Greenhouse Gas Emissions". Published online at OurWorldinData.org. Retrieved from: https://ourworldindata.org/co2-and-greenhouse-gas-emissions.

[41]   See Figure 4.5, GDP and CO2per capita, in Fletcher, William D. and Smith, Craig B., Reaching Net Zero: What it takes to solve the global climate crisis, pg. 36, Elsevier, Oxford, UK, 2020

[42]   https://www.iea.org/data-and-statistics/charts/global-coal-consumption-2000-2025

[43]   Energy Information Administration, "U.S. Electricity Generation from Renewables Surpassed Coal in April," Today in Energy, June 26, 2019, https://www.eia.gov/todayinenergy/detail.php?id=39992

[44]   See: https://www.statista.com/statistics/271823/global-crude-oil-demand/#:~:text=The%20global%20demand%20for%20crude,105%20million%20barrels%20per%20day and also see https://www.ceicdata.com/en/indicator/united-states/oil-consumption.

[45]   Strohecker, Karin, "A trillion dollar question-fossil fuel subsidies, " Reuters, November 15, 2024: https://www.reuters.com/sustainability/trillion-dollar-question-fossil-fuel-subsidies-2024-11-15/

[46]   Carbon Brief: Explainers, Q&A The Social Cost of Carbon, February 14, 2017https://www.carbonbrief.org/qa-social-cost carbon/#:~:text=Why%20do%20SCC%20estimates%20vary?%20Estimates%20of,and%20the%20way%20we%20value%20future%20damages

[47]   Pending

[48]   US energy information administration, "Planned retirements of U.S. coal-fired electric generating capacity to increase in 2025," February 25, 2025: https://www.eia.gov/todayinenergy/detail.php?id=64604

[49]   BKV energy, Lunley, Graham "Are we running out of fossil fuels?", January 2025: https://bkvenergy.com/learning-center/running-out-of-fossil-fuels

CHAPTER 6

[50]   U.S. Energy Information Agency, January 30, 2025. U.S. nuclear generators import nearly all the uranium concentrate they use. https://www.eia.gov/todayinenergy/detail.php?id=64444

[51]   Ballard, Ed, "Oil Giant Saudi Arabia Is Emerging as a Solar Power," Wall Street Journal: September 10, 2025.

[52]   NuclearNewswire, March 19, 2025. U.S. uranium production up as companies press "go" on dormant operations. https://www.ans.org/news/2025-03-19/article-6862/us-uranium-production-up-as-companies-press-go-on-dormant-operations/

53   MIT Technology Review. China's energy dominance in three charts by Casey Crownhart, July 10, 2025: https://www.technologyreview.com/2025/07/10/1119941/china-energy-dominance-three-charts/

54   Australian Government. Exporting solar energy under the sea: a potential world first for Australian technology Last updated: 04 August 2025 https://www.dcceew.gov.au/energy/publications/exporting-solar-energy-under-the-sea-a-potential-world-first-for-australian-technology

55   Our World in Data, Ritchie, Hannah and Rodes-Guirao, Lucas, "Peak global populations and other key findings from the 2024 UN world population prospects, July 11, 2024. https://ourworldindata.org/un-population-2024-revision

56   Pew Research Center, Americans' top policy priority this year? Strengthening the economy: https://www.pewresearch.org/politics/2024/02/29/americans-top-policy-priority-for-2024-strengthening-the-economy/pp_2024-02-29_policy-priorities-00_01-png/ Used with permission from Julia O'Hanlon, Pew Research Center

57   American Institute of Physics, The Discovery of Global Warming, Timeline (Milestones), February 2019, https://history.aip.org/climate/timeline.htm

CHAPTER 8

58   CarbonBrief, Rosamund Pearce, "Two degrees: The History of Climate Change's Speed limit," December 8, 2014. A graphic that shows the history of international discussions from William Nordhaus (1975) to the Cancun agreements (2010), to "hold the increase in global average temperature below 2°C." https://www.carbonbrief.org/two-degrees-the-history-of-climate-changes-speed-limit

59   Carla Tardi, "The Kyoto Protocol," Investopedia, September 26, 2019, https://www.investopedia.com/terms/k/kyoto.asp

60   Fiona Harvey, "How the UN Climate Panel Got to the 1.5°C Threshold Timeline," The Guardian, October 7, 2018, https://www.theguardian.com/environment/2018/oct/08/how-the-un-climate-panel-ipcc-got-to-15c-threshold-timeline.

61   United Nations environment program: "Emissions Gap Report 2024," October 24, 2024: https://www.unep.org/resources/emissions-gap-report-2024

62   National Renewable Energy Laboratory (NREL) Energy System Analysis 100% Clean Energy by 2035 Study. April 21, 2025

63   New codes and standards(need link)

64   NREL Energy systems analysis: 100% clean electricity by 2035 study: https://www.nrel.gov/analysis/100-percent-clean-electricity-by-2035-study

65   NREL Energy systems analysis: 100% clean electricity by 2035 study: https://www.nrel.gov/analysis/100-percent-clean-electricity-by-2035-study

66   Fletcher, William D. and Smith, Craig B., The Global Climate Crisis: What to do about it," pp. 218-219, Elsevier, Oxford U.K. 2024

67   Enerdata April 4, 2025, Global renewable capacities rose by 585 GW in 2024, with China accounting for 64%. https://www.enerdata.net/publications/daily-energy-news/global-renewable-capacities-rose-585-gw-2024-china-accounting-64.html

68   IEA: Global energy investment is set to rise to $3.3 trillion in 2025 amid economic uncertainty and energy security concerns: https://www.iea.org/news/global-energy-investment-set-to-rise-to-3-3-trillion-in-2025-amid-economic-uncertainty-and-energy-security-concerns

69   Lazard, Levelized Cost of Energy—Version 17.0, June 2024, https://www.lazard.com/media/xemfey0k/lazards-lco-eplus-june-2024-_vf.pdf/ Used with permission from Judi Mackey, Managing Director, Global Communications, Lazard.

70   Fletcher, William D. and Smith, Craig B., Reaching Net Zero: What it takes to solve the global climate crisis, pp. 204-205, Elsevier, Oxford U.K. 2020

71   Maggie McGrath, "Meet America's New Regulator: Adam Smith," Forbes, December 31, 2018, pp. 55-68, https://www.forbes.com/maggiemcgrath/2018/12/31/meet-americas-new-regulator-adam-smith/

72   RE100, "The World's Most Influential Companies, Committed to 100% Renewable Power, http://there100.org

ABOUT THE AUTHORS

WILLIAM D. FLETCHER retired as Senior Vice President at Rockwell International Corporation. After university graduation, he was an officer and engineer in the Navy, working on the design and operation of nuclear vessels. Following military service, he was an engineer with Combustion Engineering Inc., involved with the design and construction of commercial nuclear power plants. Bill also held management positions with Bechtel Corporation in Saudi Arabia planning the large Jubail industrial development project. He was a management consultant with McKinsey and Company.

After retiring, he became a member of the California Policy Center. Bill and coauthor Craig Smith wrote two previous books on climate change and global warming.

Bill has B.S. and B.A. degrees in Mechanical Engineering and in Government from Tufts University (1962) and is a graduate of the U.S. Navy's Bettis Nuclear Reactor Engineering School (1964).

CRAIG B. SMITH retired as President and Chairman, DMJM H+N, a subsidiary of AECOM Technology Corporation. Craig was an assistant professor of nuclear engineering at UCLA. After UCLA, he joined AECOM as vice president of Daniel, Mann, Johnson, and Mendenhall (DMJM). In 2000, he was named president when DMJM merged with Holmes and Narver, Inc. Craig has been broadly involved in the field of energy and power, having worked on nearly every type of electrical generating facility, including hydroelectric plants, geothermal, waste-to-energy, coal-fired, nuclear, natural gas-fired, and solar.

Craig is the author of over 100 technical papers and articles, and a dozen books.

Craig has a B.S., Electrical Engineering, Stanford University (1960) and a Ph.D., Engineering, UCLA (1965).

THANK YOU FOR READING

If you enjoyed *Irreversible, What Can We Do?* we invite you to leave a review online and share your thoughts and reactions with friends and family.

Publish Authority

www.ingramcontent.com/pod-product-compliance
Lightning Source LLC
Chambersburg PA
CBHW051411050726
47595CB00010B/4022